Gun Plumber

The Memoir of an F-105 Weapons Loader/Weapons Mechanic

by Charles L. Byler

First edition

ISBN 978-0-9980431-2-8 (Paperback edition)

ISBN 978-0-9980431-3-5 (Ebook edition)

Gun Plumber

TABLE OF CONTENTS

PREFACE

The following is an account of how I became and served as an F-105 weapons loader/ weapons mechanic, colloquially called "a gun plumber" in Air Force jargon. (The crude slang version of that is not fit for polite conversation.) That version has to do with the action of running a bore brush back and forth in the bore of a 20 mm cannon barrel. (I leave that to the reader's imagination.)

I certainly did not set out to become a "gun plumber," but a long series of events led me to that Air Force job description, and as a result of that service, to researching and writing the biography of Colonel James H. Kasler, a three-war combat veteran and the only man ever to receive three awards of the **AIR FORCE CROSS—a Decoration for Combat Valor which is second only to the MEDAL of HONOR**. There were many coincidences along the way, including either meeting or serving under any number of historically prominent individuals that I would not have otherwise.

CHAPTER ONE

Coincidence

My birth date, June 26, 1943, was coincidentally the same day Robert S. Johnson had his P-47 "Thunderbolt" shot full of holes by German fighters over war-torn Germany. His flight was escorting American bombers to Germany. Johnson spotted the German fighters first, but his radio was apparently not functioning as he called out "Bandits!" So, he instinctively reacted, turning to intercept the enemy fighters.

On a previous mission he had broken formation to meet the oncoming enemy fighters and had been severely reprimanded for doing so. Therefore, he fell back into formation—and took a pounding. His aircraft was raked repeatedly by machinegun fire, including 20 mm cannon projectiles from the noses of the German fighters. His oxygen and hydraulic systems were shot away. His engine was hit, literally blowing off cylinders. His P-47 was on fire.

Johnson tried to slide his canopy back so he could bail out of his crippled airplane, but 20 mm cannon rounds had pierced the "razor back" on his early model P-47, and torn metal would not allow the canopy to slide sufficiently rearward to allow his egress. Suffering from lack of sufficient oxygen, he struggled out of the harness straps of his parachute. He was trying to squeeze out of the cockpit—without his parachute.

As he fell to lower altitude—and denser air—he realized that bailing out without his parachute would result in certain death when he hit the ground. So, he gathered himself, sat back down in his seat, and began to assess his situation. The fire in his plane had blown out. His engine was still running, though very roughly. He discovered that his hydraulic controls were no longer functional, but that he could use trim tabs to keep his aircraft more-or-less straight and level, but he was slowly losing altitude. He also noticed that the sky around him was empty. Moments before, there had been fighters and bombers everywhere. Now, there were none. He was all alone.

His compass still worked, so he pointed the nose of his aircraft due west and headed in the general direction of England.

As bad luck would have it, he passed over a high-ranking German ace who had landed to refuel his plane and reload his ammunition. While the German fighter now had a full load of fuel, the armament men had only time to replenish the ammo in the wing guns, but not time to reload the 20 mm cannon in the nose of the fighter. That proved a life-saver for Robert S. Johnson. Compared to the machineguns in the wings, the 20mm nose cannon was a devastating airplane killer.

The German fighter ace caught up with Johnson in his crippled P-47 and began methodically raking his P-47 with machinegun fire from the 6 o'clock position on Johnson's tail. The German ace was not missing his crippled target which was unable to maneuver: a classic sitting duck. Johnson pressed against the back of his seat which had a bullet-resistant plate of steel behind it and hoped for the best.

Finally, the German ace ran out of ammunition. He pulled up alongside Johnson, saluted him, and turned back toward his air base.

Johnson flew on with his rough-running engine, eventually arriving back at his base in England. When he landed, he had no brakes and ground looped, then drifted backwards into a space between two other P-47s and rolled to a stop. To a casual observer, it might have looked like he had intentionally parked that airplane there.

Afterwards, ground crewmen tried to count the bullet holes in Johnson's P-47, but found that was impossible. Cylinders had indeed been shot off his engine. (No wonder it was running so roughly.) There were too many bullet holes in his P-47 to count, and that plane never flew again. It was simply beyond repair, but it got him home.

Another coincidence about my birth date

I sometimes joke that I was my father's "parting shot" as he went off to WWII. My father and his identical twin brother David were both drafted into WWII in the fall of 1942. Since my father was legally blind in his right eye from a birth defect, he was assigned to limited service and sent to Keesler Field in Biloxi, Mississippi as a clerk/typist. His twin brother, my Uncle Dave, had two good eyes, so he was trained as a medic and sent overseas.

Exactly nine months after my father left home for military service, I was born. Within 24 hours, my Uncle Dave's first son was born. What are the odds of identical twins having first-born sons born within 24 hours of each other? No one could have planned such an event, even if they had tried. It happened by accident. The timing event was simply that both men were drafted at the same time.

CHAPTER TWO

My Uncle Dave, Dad's twin brother, landed on D-Day at Normandy, France.

That was hell. Then it got worse. His unit was overrun by German tanks in a German forest. Everyone in the unit was killed, except for three: the unit's lieutenant, one infantryman, and my Uncle Dave (the medic).

Uncle Dave never talked much about the war, not even to his twin brother. He had medals, of which he never spoke. His sons found them on the back of a framed photograph after he died of old age. Certainly, he had a Bronze Star for service. All medics who served under fire treating wounded received the Bronze Star, but there were other medals too. He never spoke of them, never explained what he had done to earn them.

His unit followed General Patton across Europe. My Uncle Dave once said to me of General Patton: "Old Blood and Guts: His guts and our blood."

At the end of the war, Uncle Dave was kept behind as a medic to help bring American POWs out of Buchenwald. (At this point, when I tell this story, people often get an incredulous look on their faces and begin to challenge, "There were no American POWs in the death camps." Oh yes there were, and many of them died there.)

After my Air Force four-year enlistment (1962-1966), I majored in secondary English at Slippery Rock State College (later a State University), so one might imagine that I read Elie Wiesel's book *Night*, since Elie had been in Buchenwald. (There are two photographs of him in Buchenwald to prove that fact.) Actually, I read his book *Night* many years later. Elie Wiesel's descriptions of Buchenwald and what happened there exactly matched those which my Uncle Dave told me about when I returned from my 1963-1966 tour of duty at Bitburg Air Base in Germany. (My father later said that he heard more about Germany and WWII in that one session on the banks of the Neshannock Creek than he had heard from his identical twin brother in all the years from WWII until that date in 1966.) The memories were obviously simply too painful to talk about.

CHAPTER THREE

Vehicles

Fascination with a particular form of transportation vehicle seems to be a generational thing. My great grandfather, both of my grandfathers, and my father grew up with trains. My great grandfather Jonah Byler was born into the Amish faith. He split from the ultra-conservative beliefs of the Amish when he went to work on the railroad and married a Mennonite—either of which was sufficient cause for him to be shunned by the Amish community.

Before the Great Depression, my father's father (Charles J. Byler) was the foreman of a section gang, a group of maintenance men who repaired and maintained railroad tracks. My mother's father (Luther Proffit) was a conductor on the P. & L.E. (Pittsburgh and Lake Erie) Railroad, a freight line. With a lantern he waved signals to the engineer (who drove the locomotive). There were no two-way radios, walkie-talkies, or smart phones back then.

As a young man my father worked as a "gandy dancer," laying track and driving spikes to hold the rails in place.

My Uncle Andy Proffit was an engineer on the Penn Central Railroad. He gave me a Marx electric train set for my first Christmas. I know that electric train set was made prior to WWII, since metal conservation rules during the war prohibited the use of metal for making toys and the like. There were no license plates made for vehicles during 1943, the year of my birth, for the same reason. Instead, a small, square plate of sheet tin with a bolt hole in the top right corner was issued. There were also no 1943 automobiles. (Though there are rumors of a few 1943 automobiles being manufactured, in reality, those were left-over 1942 models which were kept in storage by the government long enough that they were titled as 1943 models.) Production of automobiles for civilians did not resume in Detroit until after the war was over.

I liked trains, but my real fascination was with airplanes and flying. I was a good student in school, but was not much of a pleasure reader until seventh grade. Our English teacher took my class to the library one day and showed us how to scan books, looking for something to read for pleasure. I found a Stephen Meader novel, *T-Model Tommy*. I liked old cars, and Meader made a career out of writing novels for boys. Over time, I read several of Meader's novels.

Then I found Robert S. Johnson's autobiography, *Thunderbolt*. Johnson ended up the second highest ranked P-47 "Thunderbolt" ace in the European Theater of Operation during WWII with 26 confirmed kills. (Little did I know that I would one day cross paths with Alexander Kartveli, the man who co-designed the P-47 along with de Seversky, the same man who founded Republic Aviation, who then hired Kartveli as an aeronautical engineer. (See Chapter Seventeen, pages 64 and following.)

Since I was an English Major in college, one might imagine that the first book I read by Nordhoff and Hall was *Mutiny on the Bounty*. In fact, it was not. That first book I read (in junior high school) written by those two men was *The Falcons of France* which was about their exploits flying for the French on WWI. (Years later, at a NATO tactical weapons competition in June of 1965, I saw French pilots wearing the same squadron patch of Nordhoff and Hall's unit from the First World War, an upside-down stork in flight.) After the horrors of World War I, the two aviators retired to a South Pacific Island and began writing. Some years later they wrote *Mutiny on the Bounty*, their much more famous book.

CHAPTER FOUR

Growing Up in New Castle in Western Pennsylvania

Born June 26, 1943, I was the first of five children born to Charles K. Byler and my mother, Luella Proffit Byler. We originally lived on Smithfield Street on the West side of New Castle. My sister Sally was born in 1945. Ken was born in January of 1948. Charlene was born in 1949, and Donna came along in 1952.

Speaking of births, my father was O-positive while my mother was O-negative. In that era, that was a potentially fatal combination for the unborn child. There was no treatment to inoculate against a "blue baby." The mother's immune system would see the positive baby as a foreign body and try to eliminate it like a disease. Sometimes the first baby was OK, but when a second positive baby was conceived, that baby would be killed in the womb. The doctor warned my mother of these facts.

Sally, my sister, was born without any problems. She was very bright in school and became a registered nurse. Ken came along and showed no signs of any developmental problems. By the time my sister Charlene was ready to be born, my mother was sure that she would be a stillborn baby. As luck would have it (or more coincidence) one of the attending nurses for the delivery had been a fellow schoolmate of my mother.

I must explain that natural childbirth was still years in the future at that time. Standard medical practice was to knock the mother out with anesthesia and take the baby out of the womb with a pair of forceps. My mother was so convinced that Char would be born dead that she talked to her friend, telling her that she did not want any anesthesia--telling her that she would not scream, that she would not give the delivery team any trouble. She wanted to see the baby before they took Char from her. Charlene was born a normal, intelligent child.

The doctor threw up his hands. "What do I know?" he said. This was certainly a medical anomaly. (My mother thought it was more like an act of God.)

My youngest sister, Donna, was born live—and normal, though she did seem to inherit my father's eye problem. She was dyslexic and a poor reader as a consequence. She did not like school and did not do well academically. In another odd coincidence, when I "student taught" as an English teacher, my sister Donna was one of my students—despite the fact that I was promised by the high school that would not happen.

As I was about to enter elementary school, my parents moved us to the East side of town, to 602 E. Reynolds Street (an address, like many in New Castle, which no longer exists.) Like so many houses in New Castle, it was burned by an arsonist after my parents sold it and moved on to the North Hill of town. My Grandfather Proffit's house on Shaw Street (also on the North Hill of New Castle) was likewise burned by an arsonist. It was lit off a second time before it was declared a total loss.

New Castle has many streets and avenues named after Civil War notables, such as Grant, Taylor, Long, Reynolds, Shaw, etc., besides the usual Presidents (Washington, Jefferson, Adams, Lincoln, and Garfield,) as well as schools named after other notables such as founding father Ben Franklin and Thaddeus Stevens (who was the father of free public education in Pennsylvania).

Ira Sankey, the great hymn writer, was born in New Castle (and there is still a Sankey Lane north of town). His portable reed organ is on display in the Lawrence County Historical Society located in New Castle.

I attended Thaddeus Stevens Elementary School. (It also no longer stands, replaced by a newer building.) The original was built in the 1890s—a real fire trap by later building standards. It had two stories. In the basement was a coal-fired furnace for heating. On the first floor, a wide, wooden staircase led to a wooden landing which split two ways, each then turned 90 degrees to lead to the second floor—which was also made of wood. The building was like a huge chimney made of flammable material with a live fire in the basement: an accident waiting to happen. While other such school buildings did eventually burn, Thaddeus Stevens finally fell to the wrecking ball while I was serving in the U.S. Air Force in Germany.

I was a good student, making only one mistake in my first-grade lesson book. However, I was going deaf and my first-grade female teacher, Prudence Matthews, had a high-pitched voice. I often missed spoken instructions—not because I was inattentive, but because I could not hear the teacher's high-pitched voice. I became a visual learner, of necessity—not an auditory learner. That, combined with my left-handed/right-brained thinking patterns would dictate my learning style for the rest of my life. Current teachers, please take note. (More about that in later chapters.)

When my hearing loss was identified, I was taken to an Ear, Nose, and Throat specialist. He treated me with a 10 % solution of radium on a swab of cotton—which he administered by shoving a thin metal rod (with the swab of cotton dipped in the radium solution on the tip of the thin rod) up my nose through the Eustachian tubes to dry out my inner ears. Needless to say, that was not pleasant and is not a therapy which is used today. I believe I had water in the inner ear (otitis media) caused by allergies. (My son Chad seems to have inherited that same malady and he was treated successfully by an allergist.) Today he is dual-degreed with master's degrees in both mechanical engineering and optical engineering. There is nothing wrong with his hearing or his brain.

Ironically, my Ear, Nose, and Throat specialist was a chain smoker. He developed lung cancer (surprise) and eventually died from a heart attack brought on by the lung cancer. (He should have known better.)

I might further add that my allergy problem was likely caused by a manufacturing plant which was spewing beryllium into the atmosphere about a mile upwind of our house on Reynolds Street. Today that toxic metal is not allowed to escape the confines of the building where it is used. (It can cause allergic reactions to any number of things in the environment and is ingested simply by breathing the air which is contaminated by it.)

CHAPTER FIVE

Ben Franklin Junior High, etc.

I attended Benjamin Franklin Junior High, a school which was later converted into apartments for the elderly. The other junior high school in New Castle, George Washington, was shuttered, and the two junior high schools were consolidated in a separate wing of the new high school building which was built on the site of the demolished old Ne-Ca-Hi (New Castle High) school building.

The church I attended was demolished to make way for a parking lot—part of an aggressive urban renewal project. That was a sad waste of Victorian architecture, for the city of New Castle continued to decline in population. That parking lot really wasn't needed because stores and businesses continued to close.

My parents' home on Reynolds Street (where I grew up) and my grandfather's home on Shaw Street were burned by arsonists. My elementary school and high school buildings were razed, as was my church. I literally cannot go home again.

When the Warner Brothers (yes, **THE** Warner Brothers of movie-making fame) established their first movie theater, they chose New Castle because it was the fastest-growing city in Pennsylvania. (The tin mills and other industries were booming.) The Warner Brothers (not their real name) were born with one of those hard-to-pronounce foreign-sounding Eastern European names and they lived in Youngstown, Ohio—just across the state line from New Castle. They commuted to New Castle to run their first movie theater. (The building still stands and is noted by an historical marker today.) New Castle at that time had over 100,000 residents—and was growing.

The Great Depression took a terrible toll on New Castle, which, until then, had been the "tin capitol" of Pennsylvania—as Pittsburgh had been the "steel capitol" of Pennsylvania. The tin mills failed, along with many other industries. People suffered. Times were hard. Those who could, moved on to greener pastures—some as far as the west coast.

After the Great Depression, the town fathers decided they would no longer put their trust in large industries, but would encourage smaller diversified industries. This was short-sighted as it left the economy with no backbone. One proposed steel mill rejected by New Castle's leaders, instead located in another city to the north of New Castle. For years, they had a huge sign on one of their buildings proclaiming, "This is the industry that New Castle did not want."

World War II brought some temporary relief and regeneration. Johnson Bronze made bearings for the war effort. Rockwell Spring and Axle made heavy-duty springs and parts for military trucks, and so on. I grew up in a city which seemed to constantly be in one recession after another. Eventually, the steel mills in Pittsburgh also died. Today there are no coke furnaces in Pittsburgh, and no steel mills in the city. The air there is clean and breathable.

When I graduated from high school in 1961, the population of New Castle was still above 50,000. Within a year of my graduation, over 100 of my fellow classmates had left town—with no forwarding address. (We still have not heard from most of them). More than 150 of my classmates from 1961 are now under the sod—that we know of. Many of our classmates served in Vietnam, but none died there, so far as I know—and I have checked the records.

The one Vietnam War death I have identified hits close to home. Vince Scungio was the "back seater" or "bear" or "EWO" (electronic warfare officer) who rode in the back cockpit of an F-105-G model, the rare twin-cockpit version of the F-105 "Thunderchief." The "G" model was called the "Wild Weasel" and was tasked with finding and destroying SAM (Surface to Air Missile) sites. The "G" model fired an anti-radiation missile which rode the radar beam emitted by a SAM site back to its source, thus causing the missile to go unguided. That was dangerous work, for the "G" model F-105 was pointed at the target which was targeting the F-105. It was rather analogous to two Old West gun fighters meeting face-to-face at high noon for a shootout.

Vince Scungio was the older brother of one of my fellow classmates from New Castle High, class of 1961. My brother Ken's son married a woman of the Scungio family who is the daughter of my fellow classmate. (Like I said: close to home). Years later, unprompted, the former North Vietnamese returned the remains of the front-seat pilot of that Wild Weasel, but Vince's remains were never returned. (His son, Vince, Jr., is still very angry about that). One may speculate about why it might be so that Vince's remains have never been returned. After all, both men ejected at the same time, from the same aircraft. They likely landed near each other. (It should be noted that the official record does not agree with my assessment.)

Vince, obviously, was of Italian ancestry. From all evidence, Vince was a hard-nose who likely was very uncooperative with his captors. Perhaps he was tortured and/ or beaten to death. (The trouble with bones is that they often tell forensic stories, stories which the North Vietnamese Communists did not want revealed.) Many of our American POWs who survived the "Hanoi Hilton" came home with permanent injuries inflicted in that infamous prison, or had shoot down injuries exacerbated by neglect, medical mistreatment or torture. John McCain and Vice Admiral James B. Stockdale come readily to mind, among many others. (Admiral Stockdale had his arms intentionally radially fractured so that he could never fly again, but he fooled them and did fly again.)

Such barbaric treatment was commonplace in the North Vietnamese prison system. The North Vietnamese Communists refused to admit that the Geneva Conventions for the treatment of POWs applied to American airmen. They considered our airmen to be "Yankee Air Pirates." (In defiance, one F-105 carried that slogan as nose art over North Vietnam.)

Today, the population of New Castle is 21,000 and falling. There are empty houses with no takers (at any price) on every street in town, even the formerly wealthy North Hill. Empty industrial buildings sit idle and crumbling. The town is part of the infamous Rust Belt which extends along the rivers of western Pennsylvania and eastern Ohio and up along the shores of the Great Lakes. Arson is a popular past time, so much so that New Castle made the national news a few years back for just that reason.

CHAPTER SIX

New Castle High School

In high school I took the Pre-Engineering course, a curriculum that was inspired by the Russian launch of Sputnik. That curriculum was very academic with heavy emphasis on upper-level math and science, plus mechanical drawing—done the old-fashioned way with pencil, pen, and ink. Computers were still future tense.

I decided that I would become a mechanical engineer. My plan was to go to Penn State, majoring in Mechanical Engineering. Since Penn State was originally a Federal Land Grant College (later a university), two years of R.O.T.C. (Reserve Officer Training Corps) was mandatory back then. (That requirement was ended by the anti-war Vietnam-era protests.) I willingly chose Air Force R.O.T.C.—intending to take all four years of R.O.T.C. and then be commissioned by the Air Force and become a fighter pilot. If I had succeeded in achieving that goal, I would likely not be alive today—for I would have been perfectly aligned to pilot the F-105 or the F-4 Phantom—both of which suffered high loss numbers in the Vietnam War—nearly 400 of the F-105 and even more of the F-4 (because there were more of them to lose: the Air Force, Navy, and the Marine Corps all flew the F-4 in Nam).

There should have been twice as many F-105s built, but McNamara, in one of his many bad decisions, cut the number to 833—essentially half of the original order. (Given the way the F-105s were used—or misused--against North Vietnam, we should have had that larger number.) Another of McNamara's mistakes was believing that it was impossible to hit a fast-moving jet with a manually-guided gun. He was wrong again. So very wrong.

When I was researching Col James Kasler's biography, I became suspicious that it was not MiGs and SAMs (Surface to Air Missiles) which had taken down the majority of our planes. After a time, I found the statistic: 85% of our planes (of all types) were shot down by 37mm and 57mm anti-aircraft guns—which were direct copies of the guns which Germany had used against us in WWII. (The old technology still worked, especially if we flew those planes—*as per orders*—at low level to avoid hitting civilian targets.) Stupid. Stupid. Stupid. Furthermore, our pilots were much more accurate than that. They did not need to be at tree-top level to hit targets. The best approach was a high angle dive—for both accuracy and the safety of the pilot and plane.

Many of our pilots spent long years in North Vietnam in prison under deplorable conditions of deprivation and torture. Everett Alvarez, a Navy pilot, was shot down over North Vietnam (by ground fire) on the very first mission of the war against North Vietnam. Captured, he would not be released until the war ended, 8-1/2 years later. (I sat next to him at two of the NAM-POW reunions over the years.) His book is entitled *Chained Eagle*. (Originally published by Donald I. Fine, Inc., the copyright then passed to Potomac Books, Inc., and is currently owned by University of Nebraska Press).

Foot note:

Coincidentally, I am also on my third publisher with *Tempered Steel*. Potomac Books, Inc. was the second owner of the copyright for *Tempered Steel* and the third owner of that copyright is University of Nebraska Press. When Colonel Kasler finally earned his bachelor's degree in 1963, it was awarded by the University of Nebraska. More coincidence.

Back to my high school era

I was regularly on the Honor Roll in high school and graduated in the top 10% of my high school class. I did well in math and science. However, calculus was not offered in my high school. Neither was the slide rule taught—which was the engineering computer of that era. (See *Apollo Thirteen* and watch the engineers back on earth--after the explosion aboard the space capsule--whip out their slide rules in an effort to find a solution for getting the space craft back to earth before the astronauts would run out of air to breathe.)

Those two omissions in the high school curriculum (calculus and the slide rule) helped lead to my downfall at Penn State.

I had passed chemistry in high school, but it was a near thing. My high school chemistry teacher's standard for grading centered on pop quizzes. Each such quiz involved balancing ten chemical formulas. I would find some way to make four little computational errors on each quiz. Four wrong out of ten equations was 60%. There was no partial credit. Today 60% is almost universally a "D-minus" in most school districts, but in New Castle High School back then, 60 percent was an "F." While I had started the year with "B and C" marks, and had passed three of the four quarters of chemistry, my fourth quarter grade was sliding into "F" territory. The teacher came to me with the bad news. Despite the fact that I had passed the first three quarters, the teacher told me that if I failed the fourth quarter, I would fail chemistry for the year. I was sweating bullets.

Meanwhile, if I had been my chemistry teacher's math teacher, perhaps he would have failed my course. I could do mixed fractions, something I had learned in elementary school, but my chemistry teacher had to convert mixed fractions into decimal fractions to do any computations. Really?

However, when the final exam came, it was not a teacher-made test. Rather, it was a national standard test—which used weighted questions. As a matter of further good fortune, my lab partner, Gerry, who just happened to be our class valedictorian, was away taking a scholarship test (a scholarship which he won). Had he been sitting next to me for the final exam, I am sure I would have been accused of cheating. But he was not there—and I had the highest score in the class on that test. I could not have cheated—and, fortunately, my chemistry teacher knew it. I passed, and my chemistry teacher wrote a note on my report card recognizing my "outstanding final exam" score.

However, chemistry would rear its ugly head again at Penn State. Penn State had just changed from the usual 15-week semester to the 10-week term.

The curriculum was too compressed. (Some professors, who disagreed with the change, demanded the same amount of material be covered in 10 weeks as had been included in 15 weeks—since the same credits were being awarded). Five weeks into the term we had a mid-term test. Ten weeks in, we had a final exam. Further compounding my problems was the fact that calculus and chemistry were taught in an early form of distance learning. I sat in chemistry class so far up in a theater-like arena that I could not see the facial features of the professor. He spoke to us using a microphone. For the final exam, we were told NOT to put our names on the test paper, just our seat number. (Oh great: we had been reduced to a number and stripped of our names, our identities.)

Calculus was also taught via a form of distance learning. We sat, 30 or more to a classroom, (in many such rooms) with a graduate assistant attending us as we watched our professor on a small TV screen. There was little chance for a student to ask a question or to receive help from our professor. I failed not only chemistry, but also calculus. Both were required courses for all engineering students, again taught by an early form of distance learning. Then, as now, I was not favorably impressed. I think the recent experience with distance learning during the pandemic proves my point.

Back to my right-brained, visual-style of learning

I had done well in high school in geometry (both plane and solid) and in trigonometry and, of course, in Mechanical Drawing. I could see the shapes and angles in my mind and was solving geometry and trig problems visually and then putting the solution down on paper in the way the teacher wanted to see it. But calculus and chemistry were all theory—and I had nothing to hang my hat on, nothing visual to see in my mind. I bought my slide rule after I arrived at Penn State—but, sadly, never mastered it. Sad, indeed.

There were two bright spots at Penn State: I was made a cadet warrant officer in R.O.T.C. because I could drill the other cadets in marching. (I had both marching skills and a command voice). I had been in Civil Air Patrol for several years back in New Castle. I had also marched in marching band through junior high and one year of high school.

The other bright spot was making the Air Force R.O.T.C. rifle team. I had gone through the NRA junior rifle marksmanship training program in New Castle, qualifying all the way through their highest award, Distinguished Rifleman—now labelled Distinguished Expert. My brother Ken followed in my footsteps, but unfortunately timed-out by reaching age eighteen before he had shot sufficient qualifying targets to achieve the Distinguished Rifleman rating.

The Air Force, Army, and Navy R.O.T.C. units at Penn State had an annual Inter-Service Rifle Match which was highly contested. I had the third-highest average score for the Air Force R.O.T.C. rifle team for that academic year and was on the winning Air Force team for the annual Inter-Service Rifle Match. (Not bad for a freshman.) I was in line to make the college varsity rifle team the next year.

But I knew I was done at Penn State. I was not going to be an engineer. Without a college degree of some sort, I was not going to be a pilot in the Air Force. I could not ask my parents to sacrifice any more of their hard-earned money. My father was doing hard manual labor making toilet bowls and other bathroom fixtures—lifting heavy, wet, plaster of Paris molds full of "slip" (liquid clay). The pottery building was humid and very hot from the pottery-firing kilns. (The summer I worked in that pottery, two men had heart attacks, one of whom died). The windows were all closed to keep the inferior (cheap) clay from cracking. The air was full of silica dust. (Dad eventually developed silicosis from breathing that silica dust.) My father grossed less than $4,500 per year to support a family of seven (five children and two adults.) My mother was a full-time housewife—the norm for that era. She was certainly capable of college level work, but in that era, girls were not expected, let alone encouraged, to go on to college. (Her father, raised in the hills of rural eastern Tennessee, had fled north at 18 and self-educated himself over the years into a man who was master of the *Railroad Rules* book—so much so that he could cite page and paragraph from memory like a lawyer citing case law.) Like Abe Lincoln, he was raised in a log cabin and had little formal education. However, he read every book (cover to cover) that my mother, the oldest of his three children, brought home from school, and he practiced his penmanship. He also changed the spelling of his name from "Profit" to "Proffit." I would note that there was a man named "Profit" (with one "f") on the list of passengers on the ship which arrived at Jamestown. I would also note that the area of eastern Tennessee where my grandfather was born and raised is located near the Cumberland Gap and was on the route which was traveled by pioneers who traveled west through the mountains. That route was flooded by one of the dams built as part of the TVA (Tennessee Valley Authority), a series of hydroelectric dams created to control floods and to supply electricity to rural areas of Tennessee. When I visited my great aunt's home (my grandfather's youngest sister) in the 1950s, we had to drive nineteen miles of dirt road to get to her house. Profit Hollow, where my grandfather was born, was several miles deeper into the valley from where my great aunt lived. The board house my grandfather Proffit built for his mother was still standing in the mid-1950s.

For those who like to make fun of the way hillbillies talk, I would remind that the reason they say "purdy" instead of "pretty" and "over thaar" instead of "over there" is because their sound system predates the time of Chaucer—when diphthongs (glide sounds) were first coming into use in English. Yes, their English sounds are older than those of Shakespeare—which defines the beginning of the modern age of English. Chaucer spoke Middle English. Modern day Americans would not be able to understand either the Middle English of Chaucer nor true Old English. A close cousin of Old English is Old Norse which is still spoken (and written) by the people of Iceland.

Here is an actual sample spoken by my one great uncle by marriage who grew up in that same region of Tennessee and married my great aunt—my grandfather's youngest sister.

One day when my father was visiting our relatives in Tennessee—with all five of his kids in tow, my father (who was a certified NRA marksmanship instructor) had me and my brother demonstrate our target shooting ability. Great Uncle Keck was surprised and said, "I never heered tell as how one a you tight-hided Yankees could shoot so tolerable well."

My father's response was precious. "Well, they call those 'Kentucky Long Rifles', but most of them were made in Pennsylvania." (Which is true, by the way.)

My great uncle's reference to "tight-hided" has to do with the fact that most of the men born in that generation in rural areas of Tennessee were born at home—and therefore never circumcised—thus the reference to Yankees being "tight-hided."

When my mother was growing up, career opportunities for women were very limited. A few women became schoolteachers or secretaries or nurses. Female doctors or lawyers or engineers or scientists were rare.

Mom had taken the secretarial course in high school and was literally off-the-chart in typing speed (with accuracy on the old manual typewriters). When I was ready for college, she went to work in an office job, typing and keeping track of sales for door-to-door salesmen for Jewel Tea Company. It was tedious work, balancing those many accounts to the penny.

I simply could not ask her to do that again with little hope of a better outcome. So, I dropped out of Penn State—and joined the Air Force that September.

However, that summer, I built a dirt-track stock car with the help of my younger cousin Howard and his father who was the son of a man who owned a body-and-fender repair shop. (I would note that my cousin Howard caught the bug and continued to build and race dirt track stock cars through much of his adult life.)

My father had a field car (unlicensed for the road), a '37 Buick coupe that had its body cut off (at the windshield posts and across the floor pan behind the front seat) and a wooden pickup box mounted on the rear of the chassis. I bought a rusty '38 Chevy 2-door sedan (for $10) which we also drove along the edges of the farmer's fields near the Neshannock Creek where my father built a summer cottage out of salvaged lumber. Among other things, I learned to power-slide that '38 Chevy through dirt corners—much like a dirt track racer would. Some of my fondest memories originated in and around that cottage and creek.

My '38 Chevy accidentally burned up in a grass fire while I was away at college. (Sob). Dad allowed me to take the '37 Buick, and I cut the sedan body off the '38 Chevy with a hacksaw and a cold chisel. We towed (and hauled) those two pieces into town to the body shop and welded the '38 Chevy body onto the '37 Buick chassis at the firewall. The roof rested on the roll cage, and I used sheet metal screws to attach the rest of the body to the frame rails. We spent the summer building that stock car for the local quarter-mile dirt, oval track. The body was painted a light blue metallic, a "miss-mixed" gallon from a paint shop. A number of people wanted to know the name of that shade of blue. (Sadly, it had no name and duplicating that exact color would have been neigh impossible.) The number "37" (honoring the '37 Buick which supplied the chassis and engine) was painted on both door panels.

We finished the project just in time for the last race of the summer season. That dirt-track car had plenty of torque and horsepower in the straight-eight Buick engine, a monster engine which took eight quarts of oil (with no oil filter) and whose overhead valves produced more power and speed than a flathead Ford V-8, but the coil front suspension was less than ideal for the dirt track. (Most such cars ran a straight front axle). I ran on old street tires, which were also far from ideal.

After an initial spin-out, I managed to qualify the car, but it was a coffin-on-wheels which would have passed no track's safety inspection today. I had a surplus WWII aviator's seat belt (which would be illegal today: too old), and a borrowed helmet with a half visor (which did not keep the dust out of my eyes, since I had only a window screen for a windshield). I should have had a decent pair of goggles—and a real windshield.

I held my own in the race, passing some, and being passed by others. At the end, I made a rookie mistake by taking my foot off the accelerator as soon as I crossed the finish line. A flathead Ford (which had chased me for several laps but could not pass me) smacked my rear bumper **hard**, turning my car sideways 90 degrees to my left. The engine simultaneously stalled. I sat there, helpless, as speeding race cars rushed at my exposed left side.

All I could do was grip the steering wheel and hope for the best. Gripping that steering wheel was another rookie mistake and a bad idea, since an impact on my front left wheel might have broken my hands or wrists. (Watch NASCAR drivers cross their arms over their chests once they have lost control of their car and are beginning to spin out of control). The other race cars behind me missed my car, except for one. He came across the front of my car, sheering off the front crash guard, and I saw it go spinning through the air. The front frame horns were bent sideways and the welds had failed. If I had taken a hit in the driver's door, I imagine the roll cage would have also broken apart and I would likely not have walked away from that wreck. (The professional welder had used the wrong type of welding rod.) The welds simply did not hold. My cousin Howard grabbed one of the angled pieces reinforcing the now-missing front crash guard and simply ripped it free with his bare hands. Those welds should not have failed so easily.

We towed the car home in the dark and the rain. My dad drove the tow car, and I steered the stock car (which, again, had only a screen for a windshield.) I struggled to see in the dark and the rain. I was afraid I would run into by father's car that was towing me with a chain. I had no horn to blow to warn him that I was deep in trouble trying to keep that race car from hitting the rear of his car. We parked the dirt track car on the edge of the grass near my father's cottage and I went off to join the Air Force.

CHAPTER SEVEN

The Air Force

In high school I had taken the Air Force entrance exam—just as a warm-up for the college boards. I had outstanding scores in all four test areas: mechanical, electrical, administrative, and general (three 95s and an 85). "95" was a perfect score, and one only needed a "60" to qualify for any of the four areas of choice. I chose mechanical because I knew that is where my natural aptitude lay.

When I joined the Air Force, I could have told them about my qualifications in marksmanship, my time in Civil Air Patrol (which is the official civilian auxiliary of the U.S. Air Force), and my experience as a cadet warrant officer in R.O.T.C. at Penn State. That could have started me off in the Air Force with my first promotion. Given how hard (and slowly) promotions came in the Air Force, that would have helped. However, I knew the drill about basic training. I did not want to be noticed. That could lead to being singled out for extra abuse. I wanted to remain anonymous. And I was. The training instructor didn't know my name until the last week of basic training.

However, I was one of the finalists in the selection process for cadet leader. Ultimately, the training instructor chose the best black candidate. I knew that drill too. The Air Force did not want any racists in the ranks and tried to weed such out as unadaptable. The training instructor chose a black to see if anyone would react negatively. Our Road Guards were the guys who did not know their left foot from their right foot. Our File Leaders were the "pineapples" (native Hawaiians). Again, that was for the weeding out of any racists. We had one guy named Tony who was born on Guam during the Japanese occupation of WWII. He had the distinctive Japanese vaccination scars on his upper arm, a pattern of four in a square, to prove that fact. He had enlisted in California, and he arrived in Denver, Colorado with us, and lived in the same barracks. However, he went to a different tech school—for Air Force "Ranger" training ("special ops" like the Green Berets in the Army). He went to

Vietnam, and likely survived that war. (I could not find his name listed among those killed in Viet Nam.) He was a good guy and a friend.

Recently, when I saw a feel-good Christmas movie filmed on location on Guam, I wondered why anyone would leave such an island paradise as Guam. Co-incidentally, that Air Wing on Guam is named after my old unit, 36th Wing. No men or equipment were transferred to Guam when Bitburg Air Base was closed in Germany. However, the Air Force recognized the significant history of the 36th Wing and wanted to preserve that history and memory. Astronauts Buzz Aldrin and Deke Slayton, as well as X-15 astronaut Bob White served with the 36th Wing. Double-ace and author of two best-selling books (*No Guts; No Glory* and *Check Six)* "Boots" Blesse, as well as three-war vet and Korean War "ace" Colonel James H. Kasler (also the only man to ever earn three Air Force Crosses), plus "Flying Tiger" Ace BGen Robert L. Scott (who famously penned *God Is My Co-Pilot,*) and literati James Salter (who penned *The Hunters*) had all served with 36th Wing, among many other notable pilots. As stated before, Bitburg was the premier air base in Europe.

I was given both psychological and aptitude tests upon enlisting in the Air Force. After the test results were in, I was given a list of potential specific job descriptions for which I might be selected. (Note that the *Air Force* selected: *I* did not select). At the top of the list was missile fuel systems, followed by jet engine mechanic, reciprocating engine mechanic, etc. Nowhere on the list was "weapons loader." But weapons loader fell in the category of mechanical and that was what the Air Force needed, so I was assigned. As it turns out, in retrospect, weapons loader is a job description that the Air Force has a hard time filling because weapons loaders must be on duty at all times, day and night, and in all weather conditions, outdoors.

One of our training instructors at Lowry told us that he had requested retraining at the end of his first enlistment. That request was granted, but as soon as he graduated from that tech school, he was sent back to weapons loading. Once assigned to weapon loading, one could generally not get out.

I did get one choice—between being a "461" or a "462." The "461s" were called, colloquially, "BB Stackers." They assembled weapons and transported them to the flight line. The "462s" maintained the weapons systems in the aircraft and loaded weapons onto the airplanes. I chose "462." If I had it to do over again, I would choose 461. They were not on duty "24-7, 365" like the weapons loaders, worked more inside, and carried a 1911 model .45 caliber semi-auto pistol when they were transporting nukes. I liked the .45 semi-auto and could handle it.

I knew one skinny 461 who hated the .45—with which he was forced to qualify. He couldn't hit the side of a barn if he were standing inside it. (A Texan might say, "He couldn't hit the ground with his hat.")

Ironically, that skinny "BB Stacker" had been a rated machinist in civilian life. If the Air Force had assigned him to that specialty, he would have been a tech sergeant or master sergeant in minimum time. Duh.

CHAPTER EIGHT

Lackland Air Force Base, Texas

I was inducted in Pittsburgh, Pennsylvania. Raising my right hand, I swore to "defend and protect the Constitution of the United States of America against all enemies both foreign and domestic." Note that that oath is indeterminate. That is, it remains in force for life.

We were taken to a Pittsburgh Pirates baseball game where I saw Roberto Clementi play—live and in person. (I still rank him as the greatest baseball player I have ever seen play in person.) I was saddened when he died in that plane crash—while on a volunteer mission. A brilliant career was cut short. He had excellent eye sight, so much so that he could see the spin on the ball delivered by the pitcher—and correct for it, in terms of putting the ball where he wanted it to land. As an outfielder, he would crash into a wall, if he had to, to get there in time to catch a fly ball. He could catch a fly ball far into the outfield and cut down a runner at home plate—often without a bounce. To me, he was the ultimate baseball player—with all the requisite skills. He often caught a hit ball on the fly (running in) so he could add his running speed to the ball as he threw it.

I was put on a bus with a few other recruits and taken to the airport in Pittsburgh, Pennsylvania. We flew to Texas on a civilian version of the C-54, a four- engine propeller plane.

When we landed in Texas, I thought I had walked into an oven. It was clearly over 90 degrees and very humid. Every day during my five-week stint in primary basic training, the temperature remained over 90 degrees. They flew a red flag which indicated the temperature was indeed over 90 degrees. That meant that we should not be marching nor doing calisthenics. However, we did both throughout our time there. Back then, we marched and ran in our combat boots, wearing heavy woolen socks. The Air Force was not yet issuing running shoes at that time.

One may talk about dry heat, but there was nothing dry about that heat in Texas. We sweated profusely. When we did pushups, the dust on the ground would turn into mud on our white undershirts. Our training instructor (called a "drill instructor" in the other services) made us take salt tablets. (Whose bright idea was that?) My blue web belt was salt-stained from perspiration, as were my olive drab fatigue pants. We wore a pith helmet to protect our heads from sun burn. (Our hair had been shorn off close to the scalp.) The tops of our head were white, but our lower faces were sun-burned.

One night I drew "fire watch." I should note that our barracks were WWII vintage: all wood, and therefore, a fire hazard. So, someone had to meander among the barracks of a certain area all through the night, so that an alarm could be sounded if a fire were detected. I was supposed to be relieved after about four hours, but they forgot about me, and I stood "fire watch" all night long—until the next morning. Hiding out somewhere and catching a few winks was not an option.

There was a swimming pool just a block away from our barracks. We made some noise about getting to go for a swim. Finally, the training instructor took us to the pool, and told us to get in. When we had had enough, he wouldn't let us out of the pool. We had to keep swimming—and swimming—and swimming. We never asked to go to the pool again.

We had one night march in the desert. They told us it would be ten miles. That was quite a trek. I don't know where we went or how far we actually trekked, but that march seemed to go on forever. There were stories about poisonous snakes, scorpions, Gila monsters, and big spiders out there in the desert. However, I never saw any.

Near the end of my five weeks of primary basic training, I was called to the Red Cross headquarters. I knew something was up. Sure enough: my grandfather, my father's father, had died. I was offered a chance to go home for Grandad Byler's funeral. I was told that choice would disrupt my tech school assignment, and that I would likely be sent to a different tech school, not yet determined.

I did not know, nor was I told, who would pay for my flight home from Texas to Pennsylvania for my grandfather's funeral—and for the flight back to parts unknown. I told the Red Cross man that since my granddad was already dead, there was nothing I could do for him. His wife, my grandmother, died while I was stationed in Germany. So, I missed both of their funerals. Three of my four grandparents were gone before my four-year tour of duty ended in the Air Force.

Facing an unknown future, I chose the devil which I already knew and decided to take my tech school assignment which was in Denver, Colorado.

CHAPTER NINE

Lowry Air Force Base
Denver, Colorado

Our drill instructor put a number of us on a train in San Antonio, telling us we would be on that train for 24 hours and would still be in Texas. He was right. Texas is, indeed, a very large state.

Heading north, we made a brief stop in Waco, long enough to walk across the railroad tracks into a tavern. I thought we had stepped into the Twilight Zone. The place was filled with cowboy types. It could have been 100 years ago.

Back on the train, we headed north to Dallas, where we then turned due west towards New Mexico. We stopped at Amarillo, Texas, to drop off several of our members for an Air Force tech school, and then proceeded on to New Mexico. Near the San Andreas Mountains we turned due north again—and the words of *America the Beautiful* came to mind: "For purple mountains majesty, above the fruited plains..." The mountain tops were snow-capped.

Along the way north, in Colorado, we passed by the Air Force Academy on the opposite side of a valley. The chapel on the grounds of the Academy was easy to identify because of its unique architecture.

We arrived in Denver, Colorado, for the first cold weather of the season. We were told to loosen our ties and unbutton the top button of our shirts. We were wearing our dress blues (Air Force "Class- A" uniforms) and I thought this was some kind of trick to gig us for being out of uniform. They were serious. We had risen from near sea-level to "mile-high" Denver. Such a rapid change in altitude was known to cause fainting. We needed time to adapt.

The temperature change was also noticeable. It was cold, and many of us promptly caught the flu. Lowry Air Force Base was conducting a medical study about the flu. Every time someone went to the infirmary on sick call, they would take a blood sample for analysis. They got me twice, then I just stopped going, deciding to tough it out.

We had gone through primary basic training at Lackland Air Force Base. Now we had five more weeks of secondary basic. That was mostly drill in marching, but we were finally allowed to grow our hair to a normal short length with a part.

Somehow, the Air Force figured out I had impacted wisdom teeth. An Air Force dentist pulled the three wisdom teeth that were easy to get at. However, the fourth one, in my jaw, was under the skin and lying completely horizontal. What a joy that impacted tooth extraction was! The dentist had to slice open my gum, and then he tried to drill through the tooth to make it into two separate pieces so he could extract it. He was prying with a dental tool, trying to get the tooth to break in half. My head was rocking back and forth. Finally, the whole tooth came out in one piece.

To make matters worse, they served steak for dinner that evening. Steak was so rare on the menu that I could not pass it up. However, chewing with pain killer in my jaw was not a smart idea. Several times I bit into the soft tissue of my interior cheek.

That night, when the pain killer started to wear off, I was in serious pain during tech school (which was on the 4 PM-to-midnight shift in a non-insulated metal hangar). It was freezing out. The wind was blowing. They opened the hangar doors to move an aircraft, which only increased my discomfort. The instructor recognized that I was hurting and asked me why. So, I told him about the tooth extraction, and he went to his supervisor to see if he would allow me to go back to the barracks to sleep it off. The supervisor refused to release me. More misery followed.

One of my jobs in the barracks where we lived at Lowry was fireman. One might assume that I fought fires. No, I had a card which certified me to feed coal into the furnace. The wooden barracks were WWII vintage: no insulation, and a coal furnace. I couldn't get any heat out of that coal furnace until I discovered that the firemen before me were too lazy to break up the clinkers in the grate of the furnace. Those clinkers had formed a solid mass that restricted air flow needed to keep the coal burning. I spent a while breaking up that mass so the fire could get some air. Then we had some heat.

One day, we went from shirt-sleeve temperatures in the afternoon to 20 below zero that night. They told us to "put on everything we had." They meant just that. We dressed in our olive drab fatigue pants, olive drab fatigue shirts, baseball-style olive drab caps, and our combat boots—then added our "horse blanket" (blue dress wool overcoat), plus parade hat, gloves, etc. Usually, such mixed uniforms were not allowed. The guy who was put in charge of us (another tech school student) was very tall and had prominent ears. We stood outside our barracks waiting some ten minutes for the bus to arrive. That tall young student in charge had white on the exposed upper portions of his unprotected ears. (He had only a baseball-type cotton cap on—with his ears exposed.) Sure enough, his ears were frost bitten and he had to be taken to the hospital for treatment.

In weapons mechanic/weapons loading tech school at Lowry, we learned to strip down the M-39 20mm cannon—four of which were mounted in the nose of the F-100, the fighter version of the early F-101 (but not the interceptor or the recon versions of the F-101). The gun's nickname was the "M-39 Revolver" because it had one 20 mm barrel and a large cylinder that looked like a giant cylinder meant for a huge Smith & Wesson revolver.

We also had a demonstration of stripping down the old .50 caliber machinegun because we had several men from the Idaho Air National Guard going through school with us. They were still flying the F-86 (from the Korean War era). The F-86 carried .50 caliber machineguns. (More about the F-86 when we get to the story of Colonel James Kasler.) He also fired the .50 caliber as an 18-year-old B-29 tail gunner in WWII.

The F-102 (and its follow-on, the F-106) were "all-missile" interceptors and carried no gun. The F-104 and the F-105 carried the so-called Gatling gun—a six-barrel revolving 20mm cannon. Its official name was the "M-61 Vulcan Cannon." It was dubbed the "Gatling gun" because it resembled the vintage gun of that name, but the M-61 "Vulcan Cannon" had six bolts and six barrels, whereas the original Gatling gun had only one bolt (with a spring-loaded firing pin like most other machineguns). In the F-104, the gun was powered by an external electric motor (naturally enough, since General Electric invented and manufactured it) and used conventional links similar to the old .50 caliber machinegun, though larger in size. In the F-105B model, the gun system was similar in that the ammo used conventional links and had a large drum which collected the loose empty casings. In all of the later model F-105s, the Gatling gun had a Linkless Feed System. The Gatling gun in the F-104 fired at a rate of 4,000 rounds per minute. Anything faster than that was going to break the links because the gun simply sucked up the ammo too quickly. For that reason, the Linkless Feed System was invented. The cannon shells in the later version of the Vulcan Cannon (F-105D, F, and G models) were conveyed by what looked like regular links, but those links did not grip the shell casing like conventional links. They were more analogous to a conveyor belt.

The live 20mm rounds were carried in a drum, in grooved slots with the cannon shell bases in those slots so the bullet tips pointed toward the center of the drum. As noted above, the non-gripping links acted like a conveyor belt, moving the shells through the system, through the gun, and back to the drum, where the spent rounds (or "empties") and "duds" were put back into the slots of the drum. This enabled the rate of fire to be increased to an amazing 6,000 rounds per minute—or 100 rounds per second. The gun rotated at 1,000 RPM. (It was pure mathematics, as one might expect from engineers at General Electric.) The 20mm cannon shells were electrically-primed, because the rate of fire was simply too fast for regular primers and a spring-driven firing pin. In the "Vulcan Cannon" the 20mm round was cammed into the chamber, the firing pin was cammed against the primer, and electrical voltage fired the primer. It was fast and sure.

That revolving cannon has been used ever since. It was scaled down to .308 caliber for Vietnam—and to an even smaller .223 caliber "mini gun." A three-barreled 20mm version was used on helicopters—at a much slower rate of fire. Later on, that multi-barrel cannon was up-scaled to 30mm in the A-10 "Thunderbolt 2," commonly known as the "Warthog." Talk about a tank killer. In Desert Storm, that 30mm gun destroyed over 1100 tracked armored vehicles, including T-72 Russian tanks.

But first, a short history lesson

During WWII, Alexander Kartveli was hired by Alexander de Seversky, who founded Republic Aviation. The two men co-designed the P-47. Then de Seversky was fired from his own company. (He was a disciple of Billy Mitchell, and as such, was out of step with the powers that be at that time in the Army and the Navy: thus, his firing). Alexander Kartveli became the chief project engineer at Republic. He designed the three basic versions of the F-84: the straight wing "Thunderjet," the swept wing "Thunderstreak," and the photo recon "Thunderflash." The F-84 became the U.S. Air Force's first nuclear-capable jet fighter plane).

In the 1950s, without a government contract (something no aircraft manufacturer in their right mind would do today: too expensive, too risky), Republic Aviation/ Kartveli designed the F-105 as a nuclear strike aircraft by enlarging the planform of the photo-nose F-84 and wrapping its fuselage around the shape and dimensions of the Mk-43 nuclear bomb. (That's right folks, a fighter plane with an internal bomb bay.) The F-105 is officially the "Thunderchief," but its nickname is "the Thud." Some call the F-84 Thunderflash "Thud's Mother" or the "Baby F-105." Considering the origin of the F-105, there is certainly good reason for that. It looked every bit the part.

The photo-nose F-84 "Thunderflash" was still in service with the French air force as late as 1965 (I saw one at Chaumont Air Base, France in June of 1965 when we were there for a NATO competition), and in service with the Hellenic (Greek) Air Force into the late 1990s.

Then-Colonel Robert M. White was the test pilot in the fly-off between the F-105 and the F-107 (from North American Aviation). Colonel White chose the F-105 as the better fighter plane, and North American Aviation never sold another aircraft to the US Air Force—despite their success with the P-51 and the F-100.

Colonel Bob White apparently liked the F-105 so much that he requested assignment to a frontline F-105 unit—which probably surprised the Air Force brass. He was sent to Bitburg where he eventually was one of my 22nd Squadron commanders, part of 36th Wing, at Bitburg Air Base, Germany.

Robert M. White, who retired as a two-star general, was one of my early manuscript readers of Colonel Kasler's biography in my original first-person oral history version. Reading that manuscript, I think he was horrified to learn the details of what happened to his old friend, James Kasler, during Kasler's shootdown and his more than 6-1/2 years of imprisonment in North Vietnam. (See figure 12 in the photo section of *Tempered Steel* for a group photo of James Kasler, Bob White, and James McInerny in "mess dress" uniforms of that era taken at Bitburg, Air Base, Germany in 1965—before all of them flew in Vietnam. James Kasler earned three awards of the Air Force Cross in Vietnam. Bob White and James McInerny each earned a single award of the Air Force Cross.) That is five **future** awards of the Air Force Cross, a relatively rare award—which is second only to the Medal of Honor as a combat valor award.

Midway through his safe three-year assignment in Germany, Robert M. White volunteered for the war in Vietnam where he would earn his award of the "Air Force Cross" for leading the successful (and highly dangerous) attack on the Doumer Bridge--the first successful attack on that bridge. Colonel Bob White led a flight of twelve F-105s, each laden with two 3,000-pound M118 general purpose bombs. (The F-105 had a gross take-off weight over 50,000 pounds—and could carry a heavier bomb load to target than either the B-17 or the B-24.) The F-105-D could produce over 26,000 pounds of thrust in afterburner, enough thrust to easily exceed Mach 2 in level flight.

When Republic/ Kartveli designed the A-10 (commonly known as "the Wart Hog", though its official name is "Thunderbolt II" after the P-47 "Thunderbolt"), the Vulcan Cannon was up-scaled from 20mm to 30mm. That may not sound like such a large increase, but remember that increasing the diameter of the projectile by 50% also increases the length and weight of the projectile proportionally as well as the resultant volume of propellant in the larger cartridge casing. The 30mm round for the Warthog is about **three** times as heavy as the 20mm round. Again, in Desert Storm, the 30mm "Vulcan Cannon" destroyed over 1,100 Russian tracked vehicles, including T-72 Russian tanks.

The A-10 is still flying today because there is no replacement for it. Kartveli and Republic Aviation are both long gone (dead), but Kartveli's brain-child soldiers on. (Why don't they just clone it with some updates?)

Back to the story at Lowry Air Force Base

At Lowry Air Force Base, we learned weapons loading in a hanger which held an F-100, an F-101 (the interceptor version with the rotating belly door, one side conventional and the other side of the door for the nuclear "Genie" unguided rocket), an F-102 (which doubled for the F-106 because the launchers were very similar), and a fabricated copy of the B-47 bomb bay (which doubled for the B-52.) There was no need for a complete airframe to practice weapons loading.

Across the hangar bay, in shadows at a distance, sat an F-105. A weapons instructor pointed to it and said, "That's an F-105. There aren't many in the inventory, so you are not likely to see one in the field. None of your instructors has worked on one, so there will be no instruction on the F-105."

By the way, with the "Genie" nuclear rocket carried by the F-101 interceptor version, there was no need for pin-point accuracy or a guidance system. Just fire the nuclear rocket into a formation of Russian bombers and make a hasty exit. "Close" counts in horse shoes, hand grenades, and nuclear rockets.

Since we were told by an instructor at Lowry Air Force Base that there were very few F-105s in the Air Force inventory and that we were unlikely to see one in the field, of course, I was sent to an F-105 unit: 36th Wing at Bitburg, Germany. At one time during my three-year tour at Bitburg, we had 93 examples of the F-105 ("D" and "F" models) based there in three squadrons, which is a very large number compared to the average Air Force "Wing" today.

Our air defense cover was provided by F-102 interceptors which were housed in a hangar near one end of the runway at Bitburg, and were considered a tenant unit—not part of the 36th Wing per se. They were called the 525th "Bull Dogs," and used a Mack truck bulldog as their symbol.

We were able to live-fire both the M-39 and the M-61 into a bore pit while in tech school at Lowry. What a hoot! The instructor yelled (a warning) "Fire in the hole!" Then flipped the electrical firing switch, and the Gatling gun instantly was enveloped in flames from exhausted, partially-burned gun powder (and only a 10-round burst!) Incidentally, I later heard one former F-105 pilot describe the sound of the Gatling gun firing in the F-105 as "like a dragon farting." There was no staccato like that of other machine guns---just the solid roar of 100 rounds per second of 20mm.

We also saw the live firing of a "Mighty Mouse" unguided rocket. It was almost as fast as a speeding bullet. (If you blinked, you missed it.)

There was also the detonation of a 55-gallon drum of 10% napalm in tech school. The drum was blown high into the air and landed hundreds of yards to one side—as if in slow motion. Wow!

Later, while on TDY (temporary duty) to Libya, I was able to witness the four M-39 cannons on an F-100 being harmonized (so that all four guns hit the same target point) on a bore-sight range. The F-100 vented its spent 20mm casings out the bottom of the plane. (In contrast, the F-105 retained the spent casings.) I was able to get a photo with a fast shutter speed on a 35mm camera which showed the vented casings from an F-100 stopped in the air on the way to the 55-gallon drums which collected them. The M-39 fired at about 1/4 the rate of the Gatling gun. However, with four of those guns in the F-100, the total rate of fire was equivalent to the rate of fire of the F-105—about 100 rounds per second. Too bad the F-100 went into service too late for the Korean War. It would have been a devastating MiG killer. The F-100 was our first jet fighter plane to break the sound barrier in level flight. It did that on its maiden flight. Planes such as the F-86 and MiG-15 could only break the sound barrier in a dive.

Someday, when you read the true story of the F-105 pilots who were accused of firing on a North Vietnamese patrol boat, keep in mind that the F-105 NEVER vented it spent brass. I don't know where those spent 20mm casing came from that were presented as evidence, but they were not vented F-105 rounds. (Perhaps those spent 20mm casings came from one of the many crashed F-105-D models found along "Thud Ridge"? Or, perhaps, they were collected from 20mm spent casings found on the ground from an F-100—which did vent its spent brass. (F-100s flew the Ho Chi Minh Trail in North Vietnam, etc.)

I graduated from weapons loading/ weapons mechanic school at Lowry Air Force Base in April of 1963. In an ill-advised attempt to get back to Pennsylvania cheaply, I bought a 1941 Plymouth coupe for something like $50. I bought a good used spare wheel and tire and some quarts of oil. The day I left with a passenger (who was a fellow tech school grad from one of the cities along my route to Pennsylvania) was the day the radio announced that Patsy Cline had been killed along with others in a plane crash—ironically and sadly on a return trip from the funeral for other country singers. (I still like her rendition best of the Willie Nelson composition "Crazy.") Willie wrote the song, but Patsy made it famous.

Incidentally, that '41 Plymouth did not make it very far. It began showing signs of excessive burning of oil—smoke in the exhaust--and a rather ominous rattle in the engine at highway speed. I stopped at a combination bus depot and motel in the middle of nowhere (while still in Colorado), sold the '41 Plymouth to a local rancher for $35. I and my traveling companion bought bus tickets for the rest of our trips home.

In another of my small-world coincidences, I later met a man from the local Reading, PA area where I now live who was originally from Patsy Cline's hometown, Winchester, Virginia. Coincidentally, he sang at Patsy's funeral. He also owned a restaurant in Winchester, VA, and would go into the back room after hours with her where he had a piano. Patsy could not read music, so he would play any new song on the piano that Patsy wanted to learn—so she would not have to reveal to others that she could not read music. Once again, note the coincidence.

Furthermore, I would note that neither Kate Smith nor Frank Sinatra nor "Satchmo" nor Dave Brubeck could read music (among others). That never hurt their musical careers.

CHAPTER TEN

Off to Germany for Three Years

Three years is the standard overseas tour of duty for Air Force enlisted airmen who are sent to Europe.

After my leave of absence for 30 days, it was off to Germany on a civilian version of the C-54 (a propeller aircraft) on a "brand X" airline. We were 18 hours in the air, crossing the Atlantic. There was one stopover in Prestwick, Scotland. I saw sheep grazing on the airport. Other than the grass on the airport, the terrain approaching the airport looked rugged and primitive. My return trip to the US three years later was on a 707 jet. That trip took only seven hours. We were "chasing the sun."

The next stop was Ramstein Air Base, Germany. From there, we took an Air Force blue school bus to Bitburg Air Base. The driver was an enlisted Air Force airman, but he drove like a typical European: hard and fast. I could not believe the way he rocketed through those narrow streets in German villages.

We stopped at a local "gasthaus" for lunch. The juke box was playing a pop American country song by female singer Brenda Lee. I was a little more than a bit surprised: country/pop American music in Germany! At Bitburg Air Base, I met an airman who had been in high school with Brenda Lee. Another airman at Bitburg had been a neighbor to Les Paul and Mary Ford—and had been a pal of their son. (I hope that son got one of Les Paul's guitars.) Those Les Paul guitars are valuable collector items today.

Soon I would discover "Radio Luxembourg." They were broadcasting pop music, mostly American and British tunes intended for the American G.I.s stationed in Europe, especially those in Germany. I heard the Beatles before they toured America.

More small-world: Madeleine Riffaud, the woman who owned Radio Luxembourg cropped up again, later, when I was researching Col James Kasler biography. She had visited Kasler when he was a POW in Hanoi. She was a Communist who supported the North Vietnamese in the war. Obviously, she and Kasler had very different views on that subject.

(Kasler just sat there, silent, ignoring her, while she rattled on about the virtue of the North Vietnamese cause, etc.)

I and several other new personnel arrived at Bitburg Air Base on a weekend and found most facilities closed. Therefore, we went to the base theater and watched *The Pink Panther*—twice. It was hilarious. The punch lines came so frequently and close together that I sat through the movie a second time, back-to-back, just to pick up the lines—and laughs--that I had missed the first time.

Right after I arrived, one of our Bitburg F-105 Thunderchiefs crashed on a gunnery range, fatally. It was a portent of things to come. We lost several more during the ensuing three years, plus a C-130. I eye-witnessed two of those fatal crashes. More about that later.

36th Wing at Bitburg had a long, illustrious history, dating back to before WWII. Their P-47s were the first American planes to voluntarily land on German soil. (Getting shot down did not count as voluntary.) After World War II, 36th wing was stationed in southern Germany at Furstenfeldbruck. While then-Colonel Robert L. Scott (the former "Flying Tiger" ace) was the wing commander, the wing moved north to Bitburg in the Eifel region of Germany. Scott flew into Bitburg flying an F-84 "Thunderjet" with the nose art of his "Flying Tiger" unit from WWII on the nose of his F-84 (the straight-winged version, which had a reputation as an underpowered "widow-maker.")

When James Kasler's went through gunnery training, the "Thunderjet" was used for that training in the Southwest desert. After 10 o'clock in the morning, the ambient temperature was high enough that the pilots could no longer safely take off. The hot air was simply not dense enough to sustain flight. The later swept-wing F-84 "Thunderstreak" had more power and was a much-improved aircraft. Bitburg would go on to receive the latest new jet aircraft in sequence: starting with the F-80, then the F84, F-86, F-100, F-105, F-4D, and the F-15. That streak ended with the F-15 when the Soviet Union dissolved and there were no more MiGs to intercept. The "Base Reduction and Realignment Plan" eliminated the need for the F-15 at Bitburg, and the base was closed. The 53rd "Tiger" Squadron was retired. The 22nd and 23rd Squadron were reassigned to Bitburg's sister base at nearby Spangdahlem

Air Base, but after a time, those two squadron designations also were retired.

Years later, I would meet General Scott, by accident, at Warner Robbins Air Museum in Georgia. That meeting was the impetus which led me to track down Colonel James Kasler and write his biography, *Tempered Steel.*

More about that experience in later chapters.

Bitburg

Bitburg, which means "little castle" in German, had been an ancient Celtic city even before the Romans arrived. It was originally "Beda" to the Celts and was one day's march by Roman soldiers from "Augusta Trevororum" (the "august" city of the Trevores). The French still call it "Treves" after the local tribe which originally founded it. Germans today call it "Trier"—pronounced as one syllable with a rolled German "R." The American G.I.s pronounced it as two syllables with accent on the second syllable, giving it a French twang.

Trier is arguably the oldest city in Germany and contains many Roman ruins, including part of the original protective wall around the city, as well as the "Porta Nigra," Latin for the "black gate.") It is still called that today—and no, that is not a racial slur. The sandstone structure around the gate is simply limestone which has aged to black through the centuries. (It was originally a light-colored sandstone). There are also two Roman baths, and a fighting arena where gladiators and others died, fighting to the death. In that arena is a life-size statue of Augustus Caesar. While the original marble stone which covered the original structures was long ago stripped off to decorate other buildings, the life-size statue of Augustus Caesar has never been touched. Apparently, it is an accurate image of him. He was a god to the Romans. To touch his statue would mean certain death. Trier is over 2,000 years old as a Roman city, and another 1500 or so more years before that as a Celtic seat of government. A Roman-era stone building still stands in Trier.

When I first arrived in Trier, the old "Porta Nigra" was still open to vehicle traffic. Now it accepts only pedestrians.

A huge statue of the Virgin Mary is located on a high overlook above the city on the opposite side of the Mosel River which runs by Trier. Years later, on a tour of Germany, my wife and children stayed at a hotel named (in translation) "the beautiful overlook" near that statue of the Virgin Mary.

A stone Roman bridge (Romer Brucke) in Trier dates from Roman times. There is a supposedly true story about an American soldier running through a hail of bullets, out onto that bridge during WWII to stop it from being blown up by the Germans who had set explosive charges to stop the advance of the American army.

The nearby city of Bitburg was laid to rubble by General Patton's men. (A story of revenge goes with that destruction.) The story (legend?) goes that General Paton sent a party of his men to the city to ask for their surrender. Germans killed Paton's envoys, so General Paton laid waste to the city, then cleared a path through the ruins and drove on. (I cannot verify if that story is true or not, but I have seen actual photos of Bitburg in ruins during WWII.) That much is certainly true.

There is an ancient "Romer Strasse" (Roman road or street) which still runs from Trier to Bitburg, but a wider, more modern highway is the better, straighter route today. Near the city of Luxembourg, when a modern high bridge was built, the highway engineers parked their heavy equipment and vehicles on a stretch of unused Roman road. The Romans were superior engineers. Their roadways were constructed of huge hewn blocks of stone which were buried to align with the surface of local terrain. They have never moved over the centuries and indeed are more stable and secure than modern reinforced concrete highways, and they can handle enormous weight.

CHAPTER ELEVEN

Serving at Bitburg Air Base

My first job when I arrived at Bitburg Air Base was rebuilding the 20 mm M61 "Vulcan Cannon." Most of our F-105 Thunderchiefs were manufactured in 1960 and arrived at Bitburg in 1961. By 1963 they had been through desert sand storms and several wet winters. They had been on several gunnery ranges in both Europe and North Africa, firing many rounds through their guns with no maintenance. The Gatling guns had had only emergency repairs. That was because no one had been sent to Bitburg who had been schooled in maintaining that revolving cannon. My group of new assignees was the first group which had been through such training. I was set to the task of doing that cleaning, repair, and maintenance. (I had no helper.)

All of the guns I overhauled needed new barrels. Nearly all needed new bolts and the slides that guided the bolts like tracks. Some needed new shell halves (which bolted together around the rotor like a clam shell). The rotor was the basic gun. It was likened to the frame of a revolver or the receiver of a rifle. As such, it could not be replaced. None of the guns I repaired had stopped because the rotor had jammed. In fact, the Gatling gun rarely jammed. Most often, the feed system failed, usually because of separation of the links. The one gun that needed to be "canned" was still firing. The rotor was chewed to pieces, but the gun was still rotating and firing.

At the time I arrived at Bitburg Air Base, General Curtis LeMay was the Chief of Staff of the United States Air Force. His philosophy of command and organization included the notion of consolidated aircraft maintenance. I was assigned to "36th CAMRON" (Consolidated Aircraft Maintenance Squadron). All of the aircraft maintenance personnel (including weapons loaders) were included under that designation. I wore the "36th Camron" designation above my pocket flap on one breast, and my name tag on the opposite breast. The lettering was black on a yellow background, very much like the US Army. Later on, we became "336 MMS" (Munition Maintenance Squadron—AKA "Mickey Mouse

Squadron"). Our ribbon name tags and squadron designation were changed to a red ribbon with black letters and edging. I liked that better than the CAMRON colors.

Near the end of my three-year tour at Bitburg, we weapons loaders were assigned to the separate fighter squadrons: 22nd, 23rd, and 53rd at Bitburg. I felt even better being identified that way. I was in 22nd Squadron, and proud of it. Our squadron patch, like most Air Force squadron patches, had been drawn by Walt Disney during WWII. I still have my patches and name tag from 1965, along with my patch from the NATO competition at Chaumont Air Base, France in 1965 when then-Major James Kasler was our unit commander. I wasn't allowed to keep my MA-1 jacket, but I replaced it with an identical, authentic "mil-spec" one I bought from a military surplus dealer once I got home. That jacket with its original patches is one of my prized possessions.

"Death of a Thunderchief"

During my first O.R.I. (Operational Readiness Inspection) at Bitburg, I was an eyewitness to a fatal crash of an F-105 during takeoff.

The first day of the O.R.I. was all about evaluating weapons loading on the aircraft. The second day, the pilots flew missions with dummy nukes in the bellies of their F-105s. I was in the weapons shop when I heard an explosion as one of the F-105 pilots "lit the afterburner." Standard procedure for all jet fighter planes was to run up the jet engine to full thrust, release the brakes, then ignite the afterburner once the aircraft starts to roll.

When this particular pilot "lit the burner," the afterburner blew up. I bolted out the door of the weapons shop to see what was happening. The weapons shop was located a couple hundred yards from the side of the end of the runway from which the planes were taking off. I saw pieces of the guts of the afterburner falling out the exhaust cone of the F-105. The roar of the jet engine in afterburner was abnormal: a raspy, rattling roar. Clearly, something was wrong. I had never heard such a sound before.

The F-105, already in its take-off roll, disappeared from sight behind a hangar which sat next to the flight line. The F-105 reappeared on the other side of the hangar and continued to pick up speed for take-off.

About half way down the runway, the nose wheel lifted off the runway and the F-105 began to lift for take-off. As soon as the main landing gear wheels lifted off the pavement, the main landing gear retracted—and, lacking sufficient power for take-off, the F-105 settled back onto the runway—on its wing tanks and belly. (Later, I would learn that when the landing gear started to retract, engine power was robbed by the hydraulic system for the landing gear retraction, thus decreasing thrust needed for lift-off). The two 450-gallon wing tanks scraped the concrete pavement and instantly lit off in two large puffs of flame and black smoke. With the engine still making that same rattling roar, the F-105 disappeared from sight behind a stand of trees which were located next to the runway, taxi ways, and parking areas—obscuring the F-105 from my line of sight.

Apparently, the "Thunderchief" pilot was almost flying. The nose of his plane cleared the perimeter fence, but with his tail dragging on the runway, there was no way to rotate. (All airplanes rotate about a lateral axis which runs through the wing tips. To get the nose up, the tail must drop down.) With the tail dragging on the runway, that was impossible. Deploying the tail hook to grab the wire (arresting barrier) would have been futile. In full afterburner, the tail hook would have simply been torn off. The other arresting barrier on the end of the runway was designed to grab the landing gear, which had been retracted. There was no way to stop the speeding jet plane.

Getting rid of the one-ton dummy nuclear bomb in the bomb bay would have helped lighten the load for lift-off, but that was also impossible: not enough time left to execute. For obvious reasons, a bomb mounted on the displacement gear inside the bomb bay of the F-105 could not be jettisoned by pressing the panic button (which sent jettison voltage to all the external stores pylons, but not to what was in the bomb bay.) A "SWESS" system (Special Weapons Emergency Separation System) had to be used to get rid of a nuclear weapon—even a dummy one. There simply was not time.

The F-105 was almost flying as it exited the air base. There is a farm field off the end of the runway at Bitburg which forms a shallow, sloping trough. However, as the pilot was apparently trying to nurse his

Thunderchief into the air with insufficient air speed, the right wing dropped (one wing always drops first when an aircraft reaches stall speed). The right wing grabbed earth and was torn off. I heard a loud "WHUMP"—apparently the F-105 hitting a dense stand of trees, then an echo of the impact and the whine of the jet engine's turbine blades as they were spinning down to silence.

I was soon able to walk beyond the perimeter fence at the end of the runway and could see the wing sitting upright like a billboard with the "USAF" in blue letters on the bottom side of the aluminum wing. Air policemen stopped me from proceeding farther, so I could not see the wreckage of the F-105. Just as well. (That would not have been a pretty sight.) I later saw an aerial photo of the wreckage. Only that huge J-75 jet engine was recognizable. Later, I also saw the shattered remains of the Thunderchief as it was stored in a hangar on base for evaluation.

I will never forget that smell.

Years later I met a man who flew the F-4 Phantom for the Navy. He opined that the pilot was likely killed instantly by the forces which tore that wing off the F-105. It takes a lot of "G" forces to do that to a stoutly-built fighter jet. Republic Aviation was known for its strongly-built fighter planes from the P-47 right through the A-10.

Remember when NASCAR legend Dale Earnhardt was killed when his car turned sideways and hit the wall nose-first at high speed? The human body cannot take such "G" forces. Dale's seat belt and shoulder straps held, but the weight of his head and helmet broke the bones in the base of his skull and his neck, killing him.

CHAPTER TWELVE

TDY to Wheelus Air Base, Tripoli, Libya

I was sent on several TDYs (Temporary Duties) while at Bitburg. All of the fighter pilots in Europe flew down to Wheelus Air Base, Tripoli, Libya, to do live-fire qualifications. They fired the gun at aerial and ground targets, dropped practice bomblets (small, cast- iron bomblets with only a spotting charge in the tail), fired the Mighty Mouse 3.75" folding-fin unguided rocket (with plaster-filled warheads for practice). There was a "tin can" bomblet (very light) which simulated the flight of a parachute-retarded bomb. The Sidewinder heat-seeking missile was simulated. The pilots did not fire the command-guided AGM-12B, the so-called "Bull Pup." (In actual combat, they flew the "Bull Pup" with a joystick in their left hand, while they flew the aircraft with their right hand.) In Vietnam, the "Bull Pup" proved to be of insufficient explosive power to knock down bridges, etc.

We also practiced nuclear delivery. Usually, that was done with the same 25-pound cast iron bomblet that we used for conventional bombing or the "tin can" bomblet we used for "chute-retarded bombing." However, there was a full-sized inert version of the MK-28 nuke (called a "Shape") which we loaded only very rarely (because it was so expensive.) That led to another story of coincidence.

In New Castle while I was in high school, most people knew that Lockley Machine Company in New Castle had a secret government contract. But, of course, because it was "secret" no one seemed to know what that was.

One day when I was on TDY to Wheelus Air Base, Tripoli, Libya, one of those rare "Shape" dummy nuclear bombs was delivered to the flight line. On the tail section of that dummy bomb was a decal: "Lockley Machine Company, New Castle, PA." Mystery solved—and another small-world story added to my collection.

In fact, the exact dimensions (shape) and weight of that Mk-28 bomb were classified. Decades after I left the Air Force, I was not permitted to even say "nuclear bomb" or Mk-28 or Mk-43. The F-105 has been out of the Air Force inventory since 1984, and all of the nukes from that era have been de-activated and their precious metals and radioactive materials salvaged.

Some years later, back in civilian life, I made a presentation about Colonel James Kasler's biography *Tempered Steel* at the Historical Society in New Castle. In the first row sat a former Navy Blue Angel.

Also in attendance was a man from Mooney Brothers Supply Company in New Castle who had filled the "shapes" with concrete to bring the bomb to the exact weight of the real thing. He had also been a reserve B-47 pilot who had practice-dropped the "Shape." He was literally one of the Mooney Brothers. His brother Ralph Mooney had been my Sunday-School teacher when I was in high school. Once again, it truly is a small world.

Napalm was in the inventory back then, (since banned) but I only loaded that "live" one time—at Bitburg for a NATO "air power" static display—which included an F-102, an F-105, and a French "Mirage" sitting side by side by side on a ramp at Bitburg.

NATO equipment was supposed to be compatible with US made equipment, but an incident that day of the loading proved otherwise. After we had de-armed the napalm tank and down-loaded it, the French pilot tried to start the jet engine of his "Mirage." An American-made high pressure air compressor had been hooked up to the jet engine of the "Mirage" (the same ground power unit commonly used to start the F-105 and other US jets.) But several attempts failed to produce enough RPM to "light" the fire in the French jet. I could see that the French pilot was getting frustrated. (Perhaps he had a hot babe waiting for him back in France.) Finally, he "lit the fire," but, by that time, jet fuel had puddled in the engine, and had dripped onto the concrete parking pad.

When the French pilot "lit the fire," a fireball fell out the exhaust cone of the "Mirage" and flames started following the drip from the tail cone toward the jet fuel puddled in the jet's tail. Myself and the rest of the loading crew decided that it was high time to get out of there before things got even hairier.

Meanwhile, the crash rescue crew in their huge Le France fire-fighting truck had been day-dreaming on the job, sitting inattentive in the cab of the fire truck. Finally, one of them spotted the fire and the ensuing larger problem. They literally spilled out of the high cab of the fire truck from both sides—onto the pavement like a bunch of circus clowns. Then they scrambled around like comic Keystone Cops, trying to deploy their hoses to fight the fire.

They were able to douse the fire, but that French jet did not leave Bitburg that day. The fire had done damage. Somehow, an American-made high-pressure hose and fittings measured in inches did not mate up properly to a receptacle on a French manufactured jet aircraft made in France to a metric standards. Fortunately, our superiors discovered that fact before we tried to go to war against Russian jets with a mix of French-made jets and American-made "ground power" equipment.

CHAPTER THIRTEEN

X-15 Astronaut Colonel Bob White, My 22nd Squadron Commander at Bitburg

As I noted before, one of my 22nd Squadron commanders at Bitburg was none other than former X-15 astronaut then-Colonel Robert M. White, who would later retire as a two-star general. He had been the first man out of the atmosphere in a rocket plane (the X-15). He was also the first man to exceed Mach 4, Mach 5, and Mach 6. (Again, in the X-15).

One might wonder why Robert M. White did not get more acclaim at the time for his altitude and speed records in the X-15. That was a matter of security and the nature of the mission plan for the X-15. Nowhere in that mission plan was breaking speed and altitude records. The sole purpose of the X-15 was to prove the efficacy of the nickel-titanium hull of the X-15.

Fact: at about Mach three, aluminum begins to soften because of heat friction from the air. The Air Force was looking ahead to the SR-71, which was intended to exceed Mach three—and then some. The hull of the X-15 was milled out of a single billet of that nickel- titanium alloy. (There was no way to stamp out panels of such a hard alloy.) At that time there was no defined altitude for the limit of the atmosphere. (There was no sign up there stating "You are now leaving the atmosphere.")

The actual top speed of the Sr-71 was never revealed. However, on the last official flight of the SR-71, one flew coast-to-coast in under one hour. Do the math.

Bob White was also the test pilot who selected the Republic F-105 over the North American F-107. White chose the F-105 as the better fighter plane. Both planes were capable of Mach 2+ because they shared the same engine, a J-75. The F-107 was essentially a highly modified F-100 with a pointy nose and the jet engine intake above and behind the cockpit instead of an open shark's mouth in the nose. But, the F-107 had a detachable bomb bay, whereas the F-105 had an integral, internal bomb bay.

After his test-pilot days, Bob White requested assignment to a front-line F-105 unit. Apparently, he really liked the F-105. They sent him to Bitburg, which, again, was regarded as the premier fighter wing in Europe at the time. Its history was replete with astronauts and aces, plus many who made general.

Sometimes when I am asked what I did for three years at Bitburg, I answer, "I helped make two-and-a-half generals." A colonel from the Pentagon, who had potential but not the necessary command experience, would be sent to Bitburg. He would wring out the wing in preparation for an "Operational Readiness Inspection." If the wing passed that O.R.I., he would get his promotion to Brigadier General and move on to higher command.

Part way through a safe European three-year tour of duty, Robert M. White volunteered for Vietnam—which was certainly NOT a safe assignment. (As noted earlier, more than 300 F-105s were lost in South East Asia.) In the Vietnam War, White earned an award of the Air Force Cross for leading a flight of twelve F-105s, each laden with a pair of 3,000-pound M118 general purpose bombs against the Doumer Bridge. (That's a 6,000- pound bomb load on a single-engine fighter plane—which is a much heavier bomb load than a B-17 or a B-24 carried to Germany in WWII.) Gross take-off weight of the F-105 was over 50,000 pounds. Its jet engine produced over 26,000 pounds of thrust in after-burner. In Vietnam, White flew 70 combat missions and never collected a single bullet hole. In WWII, he was hit by a single bullet in the cooling system of his P-51 "Mustang" while strafing a German airfield near the end of WWII—and briefly ended up as a Prisoner of War. (Compare this story with the one about Robert S. Johnson's P-47 which was shot full of more bullet holes than could be counted, yet that P-47 carried Johnson all the way from Germany back to his base in England.) I rest my case in the argument about the P-47 versus the P-51. Yes, the P-51 was faster and sexier looking, but the P-47 could take much more pounding than the P-51 and still get its pilot home.

Later, Bob White would command "Fourth Allied Tactical Air Force," a NATO designation, and would be promoted to Brigadier General. As stated before, he retired a "two-star" general.

Bob White's biographer was his golfing partner in Florida—who was a published writer but knew nothing of Bob White's flying career: not his World War II experiences, not his Vietnam experiences, not his Collier or Harmon Trophies, nor his other test pilot awards or records. He didn't know about White's Air Force Cross award for valor. (I have often wished that I could have interviewed Bob White about the Doumer Bridge Raid). I would have (politely) asked him if he "violated orders" when he turned his flight of twelve F-105s (in "full afterburner") and flew directly over Hanoi to announce their presence to the American POWs in the "Hanoi Hilton." (That was a morale booster for the POWs.)

If he had followed orders, he would have turned down river—right into the path of waiting anti-aircraft gunners. It is highly likely that there would have been losses, and Bob White was leading them.

At the 25th year reunion of the NAM-POWs (Prisoners of War from Vietnam) in Dallas, Texas, I sat at table for dinner with Colonel Kasler (as his guest) and a man who had been in the planning room in Vietnam during the war. That colonel told me that when the Vietnamese "help" came in to clean the planning room, his men would cover the mission boards (planning maps) with butcher paper to conceal the plans.

"Hell," he said. "We should have left those boards uncovered. The hired help could have updated them for us!"

In truth, the enemy always seemed to know when the American forces were coming—right down to their flight paths going to and exiting from the targets.

It seems that our enemy in Vietnam also had spies even in our Embassy in Saigon. In another interview (with a member of the Marine Guard assigned to protect the Embassy in Saigon—during the Tet Offensive), that Marine stated that two of the embassy drivers were double agents who gunned down the guards at the back gate of the Embassy (by shooting them in their backs) when they turned toward a decoy explosion at the front of the Embassy. That Marine I interviewed would later kill a Vietnamese employee of the embassy with whom he had played cards in the Embassy. That Vietnamese employee was responsible for killing a Marine guard who died during the assault on the Embassy, which is called "the battle for Bunker's bunker." The actual Embassy building was never breached.

In a typical jurisdictional dispute between agencies, the Marines and the Embassy each insisted that the other was responsible for arming the Marine Guards. The Marine guards had Embassy submachineguns while on duty at the Embassy. The Marines had only revolvers for proficiency target shooting practice at their residence location. So, when the Embassy was hit, the reinforcing Marines (from the Marine House) carried only .38 caliber revolvers with low velocity target ammo. However, those Marine reinforcements, along with CIA types in civilian clothing and the on-duty Marine guards, saved the day. The Vietnamese police, who had a station just down the street from the Embassy, never lifted a hand to engage the attackers.

("So it goes," as Kurt Vonnegut wrote so often in *Slaughter House Five.*)

Coincidentally, both James Kasler and Kurt Vonnegut graduated from the same high school in Indianapolis, Short Ridge High. However, they never met because they were just far enough apart in age.

Another of my 22nd Squadron commanders at Bitburg Air Base was George O. Guss. The F-105-G model he flew in Vietnam is in the Air Force Museum in Dayton, Ohio. Its tail number is 320 and has three MIG stars painted on its nose section. George O. Gus was responsible for one of those MiG stars, though the US government officially recognizes only "half of one kill" (shared with a Sidewinder missile fired by an F-4 Phantom jet.) It seems like the US government was unwilling to reveal that we were shooting down MiGs. (Why?) On one of Guss's early combat missions in Viet Nam, his flight of "Wild Weasel" F-105s was attacked by MiGs. To lighten the load for combat engagement, Guss hit the panic button which sends jettison voltage to all the external stations (wings and centerline.) That punched off the pylon which carried an anti-radiation missile intended to kill SAM sites, plus the anti-aircraft missiles used for self defense, as well as an external fuel tank. Guss's wing mates saw him fly into a cloud formation with a MiG on his tail. There was an explosion in the clouds, and Guss's wing mates thought that Guss had been hit by the MiG and had blown up. However, his F-105 emerged from the other side of the clouds, while the MiG did not. Apparently, the MiG had flown into the pylons and all those heavy missiles which Guss had jettisoned from his F-105—along with any jet fuel carried externally. However, his MiG claim was denied—for lack of gun camera film documentation. (Of course, there was no gun camera film: the gun camera records what happens when the gun is fired **FORWARDS**). There is no gun camera looking aft.

Years later, always suspicious that the US Mig claims by US pilots were under counted, I found one possible explanation: President Johnson had blown his top and shouted, "I don't want to hear of any more MiGs being shot down!" (He was foolishly trying to negotiate a settlement with North Vietnam so US forces could disengage and go home.)

CHAPTER FOURTEEN

One Hairy Loading Experience at Bitburg

I had several weapons loading crew chiefs at Bitburg. One was a stutterer. (He didn't last beyond the first test-loading evaluation.) The man shouting the commands while loading weapons was not permitted to stutter—for obvious reasons. ("B-B-B Bomb b-b-bay d-d-d doors open!")

Another weapons loading crew chief did not stutter, but he would not have qualified for the "Einstein Award." He spent most of his time brown nosing the tech sergeant—looking for his next promotion. (He literally spent his time following that tech sergeant around like a puppy following his owner—right on his heels.) Meanwhile, we other three on the weapons loading crew did all of the work. He was also the guy who pushed the wrong button in the following incident.

We were in a phase where, for whatever reason, we were reconfiguring the F-105s each day from nuclear to conventional readiness. For conventional, we added outboard pylons to mount an extra conventional bomb on each outboard wing station.

For nuclear, we dismounted those outboard pylons and readied the bomb bay to accept a nuclear bomb. With so many planes to reconfigure each day, we went off-checklist and multi-tasked to save time. "Tiny" (an ironic nickname: he was both large and muscular) our MJ-1 bomb lift driver, had positioned the MJ-1 lift arms under the "displacement gear" (a nitrogen charged bomb kicker) in the bomb bay (which meant that the engine and seat of his MJ-1 were parked under the right inboard drop tank.)

I was up on the right wing as a wingman. I had just torqued down the spanner on top of the wing to secure the outboard pylon, then had come down to the ground to check the continuity of the electrical connection of that outboard pylon with a PSM-6 multi-meter. I did not get a positive response. A ground-power unit was running right next to the F-105 to provide electrical power to the aircraft without starting the jet engine. It was noisy, so I signaled to my weapons loading crew chief to send jettison voltage to the outboard pylon stations. (I would note that we never installed jettison cartridges in the outboard pylons because they were so light that there was no need to jettison them in an emergency.) I figured that my weapons loading crew chief (a staff sergeant) would push the "outboard-station" jettison button, since he was looking out the right side of the cockpit and those selective jettison buttons were on the right side of the cockpit. But, instead, he turned to his left toward the panic button which sent jettison voltage to *all* of the external stations.

I said "BANG" in my head just as he hit that panic button—and the right-wing drop-tank instantly was jettisoned right on top of the engine and driver's seat of the MJ-1—and the damaged drop tank began leaking jet fuel. I just shook my head in disbelief, turned, and walked away. If Tiny, our MJ-1 driver, had been sitting on his exposed driver's seat, he would have been crushed to death by the fuel tank. However, he was off that seat and in the bomb bay area adjusting the lift arms of the MJ-1.

If the fuel cell had ruptured more than it did, the hot, air-cooled engine of the MJ-1 would likely have ignited the jet fuel. The resultant fire could have destroyed an expensive jet (about two million dollars each back then), and there might well have been injuries and/or deaths. Also, since the F-105s were parked wing tip to wing tip, there could have been multiple losses of F-105s.

Why did that drop tank jettison? After all, there was a safety pin in place on that drop tank's pylon. It is usually a fact that several things have to go wrong to produce an accident. We found later that that outboard pylon had been written up as faulty. However, that write-up form had slipped down inside the pylon and no repair was made. (There was a faulty electrical connection inside the pylon.) Secondly, the weapons loading Crew Chief should have known to use the selective jettison button (inboard, outboard, and centerline) instead of the panic button. Third, the safety pin system in the drop tank pylon was faulty. Fourth, we were off-checklist with our procedure. We were just plain lucky—or, perhaps, this was one more example of divine intervention. Someone (or several of us) could have been killed.

Then our load crew chief had the nerve to ask me and the other members of the load crew to contribute to the $50 fee the Air Force charged him for the lost jet fuel. I just turned and walked away without comment. The Air Force considered the drop tank an expendable item and did not assess any charge for its loss. Actually, that tank was an expensive item compared to the value of the lost jet fuel.

I exhibited a startle reaction to sudden, loud noises for a long time afterwards.

CHAPTER FIFTEEN

Chaumont Air Base, France

In June of 1965 I had a different and very competent weapons loading crew chief whose real first name was "Crewson." (Get it? "Crewson," a "crew" chief). The other members of that load crew were also very competent. As proof of that, our loading crew was assigned the role of "standardization team" for weapons loading for our whole squadron. That meant that we were responsible for training other weapons loading teams and certifying them. Because of that designation, we were chosen in June of 1965 to represent the entire United States Air Force in Europe at a NATO AIRCENT tactical weapons competition at Chaumont Air Base, France. The F-105s from Bitburg and our sister base Spangdahlem represented USAFE (the United State Air Force in Europe) for that competition.

Then-Major James H. Kasler was chosen as our detachment commander for that event. He was one of our operations officers at Bitburg Air Base which meant that he was a "mission planner." At Chaumont, France, he was selected as the overall commander of the "Fourth Allied Tactical Air Force" unit for that event. (*Tempered Steel* does not identify the fact that Kasler was the overall commander of the "4th Allied Tactical Air Force" unit in that NATO competition because I could find no confirmation of it anywhere at the time we went to press with *Tempered Steel*). Apparently, Kasler was selected too late to get his name in the printed program for that event. Nonetheless, it is a fact that he was the overall "Commander of the Fourth Allied Tactical Air Force" team for that event. We were pitted against the "Second Allied Tactical Air Force." The NATO air forces in Europe were divided equally between the two Allied Air Forces for that competition.

The Canadians were flying their own-built version of the F-104. The Germans were flying American-made F-104s. The German F-104s were split between the two NATO-designated Air Forces. The French were flying American-made F-100s. The English were flying their "Canberra" two-man bomber. It was sub sonic, and with a crew of two men, should have been superior to single-seat fighter planes in the bombing competition. However, the much faster F-105 out-scored the Canberra. Two of the NATO countries were flying the old F-84 "Thunder Streaks," which were museum pieces even then.

The contrast between the F-104s in the two German units and the Canadian unit was stunning. The Canadian F-104s looked like shiny new silver dollars and were well maintained. Their ground crewmen wore immaculate white coveralls. The F-104s in the two German units looked like crap. They were painted in drab camo colors and looked like they needed to be washed, among other things. Their maintenance men were not in service long enough to become proficient. Their idea of checking out an airplane was to fix it, then turn it over to a pilot to take up in the air to see if it worked. The concept of "ground checking" totally escaped them. The Germans crashed two of their American-made F-104s during the two weeks of the NATO competition.

The first loss happened before the event officially began. One German F-104 pilot took his aircraft out over the bombing range for a familiarization pass. When he extended his speed brakes to slow down for a closer look, the engine flamed out. He ejected and landed safely less than 100 yards from the burning wreckage of his aircraft. (I have no idea what caused that crash. Was it a mechanical problem or just pilot error?)

I was an eyewitness to the second crash. I was standing parallel to the approach end of the runway when I heard a loud POW!! I jerked my head towards the runway in time to see that the F-104 pilot had ejected. His chute immediately popped open. He swung like a pendulum, feet-forward, then feet-aft, then landed on his feet and began reeling in his parachute calmly as his F-104 continued to land itself and then rolled down the runway—before drifting to the right and running off the runway—which sheared off his nose gear and right main gear. The F-104 went nose-down into a large drainage ditch next to the runway and sat there sucking dirt and dust into the intakes and blowing it out the tail pipe. Someone from "crash rescue" knew how to shut down the jet engine and did so. Later, we heard that this was not the first F-104 that particular German pilot had ejected from. Apparently, he was afraid of the F-104 or simply did not trust it.

I watched a weapons loader Frenchman load folding-fin rockets into the launch tubes of an F-100—without first doing a stray voltage check for static electricity. That was a clear violation of *our* safety procedures for loading a very sensitive electrically-launched missile. Hairy stuff.

Meanwhile, General Curtis LeMay's "consolidated maintenance" philosophy kept me from being promoted for a long, long time. I was up against 3,000 other mechanics (with various specialties) on the promotion list at Bitburg. Never mind that weapons loading was critical to our "nuclear deterrence" mission. Nearly two years passed before I received a second stripe, and my third stripe was essentially a going-away present at the end of my three years at Bitburg, and about 90 days before my four-year enlistment ended. Too little. Too late. People sometimes ask me why I did not reenlist. First of all, the pay scale back then was little improved over that of WWII servicemen. I had to get past two years of service before my monthly paycheck hit $100. My best twelve months in the Air Force did not gross $2,000. The re-enlistment bonus would have been under $4,000 for four years. That's less than $1,000 per year. I figured I would be much better off going back to college to earn a degree.

It was a matter of good timing (another coincidence) that the Vietnam-era G.I. Bill for college education benefits was passed the spring I came out of the Air Force. I was given an "early out" (from my four-year contract) because I would have had so little time left that it would not have been worth the government's money to send me to my next assignment. The Air Force also owed me 60 days of back leave (because getting leave as a weapons loader was next to impossible.) We had a master sergeant in charge (NCOIC) who had a twisted sense of humor. He delighted in tearing up an application for leave in front of the applicant. (He had the right to deny a leave, but he had no right to tear up that application for a leave—let alone in front of the applicant.)

Furthermore, my next duty station had to be an F-105 base in the states. (One could not serve two consecutive overseas assignments—and I was "prefixed" for the F-105.) There were only three F-105 bases in the states. Two of the three bases were rotational to Vietnam. The third base, Nellis, near Las Vegas, was a training base for fighter pilots. It was in the hot desert where weapons loaders had to go hatless (to prevent the hat from being sucked into a jet engine) and often shirtless (because of the heat). A weapons loader would be "humping weapons" all day, all the time. (I was 21, so I could get a "class A pass" to go into Las Vegas, but I wanted no part of "Sin City.") First of all, I am not a gambler. Among other things, I still did not drink—a rarity among weapons loaders.

In Germany I had an occasional glass of Mosel white wine when I treated myself to a dinner out in the town of Bitburg at a modest restaurant. (I was often mistaken for a German native when I went to town dressed in my green jacket, a cravat, dress slacks, brown shoes, and my German green fedora—complete with a game bird feather and a couple of pin decorations which included a miniature October Fest mug.) I could order a rump steak, smothered in mushrooms, with a side dish for the equivalent of $1.25 US. A glass of Mosel white wine was the equivalent of 25 cents extra. (The German mark was worth almost exactly one US quarter back then): four German marks to the dollar. I could have walked into any bank or most jewelry stores and bought common-date US gold for $35 an ounce. Ah, the good old days.

I was told (incorrectly) that it was illegal to import US gold coins into the US. Since I have been a coin collector all my life, I smuggled one US gold piece through customs—an 1853 one-dollar gold coin, the most common date US gold dollar, and therefore the least valuable. I later gave it to my daughter as a gift. I hope she still has it. Any gold coin is worth a whole lot more than $35 an ounce today.

CHAPTER SIXTEEN

James Clark

1965 was a banner year for me: a photo I took of Formula One race car driver James Clark racing at Zandvoort, the Netherlands, won first prize in a photo contest in the "sports or action" category for all of USAFE (United State Air Force in Europe) that year.

I took that photo with an inexpensive "Petri 7-S" range-finder 35mm camera, my first 35mm camera. Since that model would not accept interchangeable lenses, I had no option of a telephoto lens. I had to get close to the racecourse—which was allowed back then. Also, my fastest shutter speed was $1/500^{th}$ of a second, not the faster $1/1000^{th}$ of a second of most single-lens reflex cameras. Therefore, I had to pan the race cars as they sped by on a road course. The results could not have been better. A faster shutter speed would have produced a photo that made the race car look like it was parked, sitting still. $1/500^{th}$ second shutter speed produced just the right amount of blur to the wheels and the streaks of green grass and yellow wildflowers in the background—while the race car body was not blurry.

Not only did James Clark win that Dutch Grand Prix Formula One race, but he also won the next Grand Prix Formula One race—which was held at the Nurburg Ring in Germany. That racecourse just happened to be only 50 miles from Bitburg Air Base. I was there. Clark led every single lap from start to finish. That win in Germany sealed the Formula One World Champion title for Clark in 1965. The next year, Lotus (for whom Clark drove) did not have a car to enter in the changed Formula One. Therefore, he could not defend his title.

Since Clark was contracted to drive for Lotus, he switched to racing their Formula Two and Three. Disaster struck when one of those Lotus cars broke on him, sending him into a forest of trees, killing him. Clark did not make a mistake. The car simply broke. Or maybe a tire was going down. In the aftermath of the crash, it was impossible to determine which part might have broken before the crash and which parts were broken by the crash. The experts agree that Clark could not have made a mistake.

His driving skills were such that watching him drive was like watching the movements of fine Swiss watch. Lap after lap was precisely the same on each corner.

Should you think that the Nurburg Ring is not a tough course, racing legend Sterling Moss stated that it was the only course he never fully memorized—with over 1000 hills, dips, and curves in every lap. He memorized ALL the other courses he ever raced on.

Years later, Niki Lauda crashed on the Nurburg Ring, knocking him unconscious. His car caught fire—and burned the whole top of his head, leaving him with hideous permanent scars for the rest of his life.

I received a letter of commendation from Colonel George O. Guss, who was my Squadron commander at Bitburg at that time, congratulating me on winning that photo contest. (See my previous musings about his F-105, number 320 which is in the Air Force Museum with three MiG stars painted on the nose, and the story of his MiG claim, which was denied for lack of gun camera film evidence.)

CHAPTER SEVENTEEN

Alexander Kartvelli

In the latter part of my three-year tour of duty at Bitburg, as noted before, my weapons loading crew was designated as the "weapons loading standardization team" for our squadron, which meant that we were responsible for training and certifying the other weapons loading crews. As a consequence, we had been sent to the NATO weapons competition at Chaumont, France, serving under then-Major James Kasler. That fact was critical to what happened later. More to follow about Colonel Kasler in later chapters.

One day, we were sent to the flight line and introduced to "an engineer from Republic Aviation." No name was given, but I was a bit puzzled by this individual. He was an older man, but he was not dressed like an American engineer or businessman. Rather, his manner of dress was like that of an older European man. He wore a dark-color heavy overcoat, a conservative dark suit, polished dark dress shoes, and a dark "Dick Tracy"-style Fedora hat. He walked with a noticeable limp. I thought immediately of Alexander de Seversky because he had been injured in a test plane crash which left him with a limp, but he had been forced out of Republic Aviation during WWII. This man couldn't be Seversky. When this "engineer from Republic" talked, (which was but little) it sounded to my untrained ear like a Russian accent.

We did the test loading of a dummy Mk-57 nuclear bomb. The Mk-57 was to be the next nuclear bomb for the F-105 in the "nuclear deterrence" role in Europe. However, since the F-105 was literally designed around the shape and dimensions of the Mk-43 nuclear bomb, the Mk-57 was too long to fit inside the F-105 bomb bay. So, a new centerline pylon was designed at Republic Aviation to fit on the closed bomb bay doors to carry the Mk-57 on the external centerline station of the F-105. We were to do the test loading for the "engineer from Republic" (to make sure everything fit and that the weapons loading check list was functional.)

The test loading went off without a hitch of any sort, and the "engineer from Republic" got a little smile of satisfaction on his face.

Years later, when I was doing research related to Colonel James Kasler's biography, I happened to be researching Alexander Kartveli on the Internet. Usually, I had seen only photos of him as a younger man. Now I came across a series of photos, including some of him as an older man. Bingo. There was the "engineer from Republic" for whom we had done the test loading. The engineer from Republic was not just any engineer: he was the Chief Project Engineer, Alexander Kartvelli himself, the man who had co-designed the P-47, and had designed all the variants of the F-84, also the F-105, and was still at Republic for the development of the A-10 "Thunderbolt II" (better known as "the Wart Hog")

Since I had interviewed Major General Hoyt S. Vandenberg, Jr. for Colonel James Kasler's biography, I emailed him and asked Vandenburg, Jr. if the "engineer from Republic" could possibly have been Alexander Kartveli himself. He replied, "Oh sure. He was a frequent visitor to Bitburg."

That made sense. There were only seven wings of F-105s worldwide: three in the States (which had no nuclear-deterrence role since there were no Russian targets within range), two in the Far East, and two in Europe. The Far East was much farther in miles from Republic's headquarters on Long Island than the two bases in Europe. Of those two bases, Bitburg was the "premier" base. So, of course Kartveli would choose Bitburg as the F-105 base to visit.

The A-10 Attack Plane

As noted earlier, the A-10 destroyed over 1100 tracked vehicles during Desert Storm and could do likewise to Russian tanks in Ukraine. A-10s could also have wiped out forty miles of Russian supply trucks and stalled the Russian offensive, early on. I am sure that Ukrainian pilots who can fly sophisticated MiGs could adapt easily to the A-10. It's a slower and more forgiving plane that can fly on one engine and be landed with the wheels up. (The landing wheels are partially exposed when the landing gear is retracted.) The A-10, if it loses hydraulic power, gives the pilot the option of locking out the hydraulic system so that the plane can be flown manually. That can be a life saver in the event of combat damage to the hydraulic system. Though it is not recommended, a few pilots have landed the A-10 with the hydraulics locked out. Of course, that means that the pilot has no brakes or steering, but it has been done successfully, nonetheless.

If James Kasler had had such a lock-out system in his F-105 when he was hit by 57 mm cannon shells over North Vietnam, he might have been able to maintain control of his F-105, at least long enough to get out of the immediate area of his shoot-down site. At minimum, such a system would have avoided his horrendously shattered leg which was caused by the loss of hydraulic control of his stick - which locked his right knee against the right side of his cockpit. Firing his ejection seat then shattered his thigh bone, and rammed the splintered bone up through his groin area into his gut. (Ouch!)

There is one further observation here. Kartveli's native tongue in what became Soviet Georgia is a close cousin of Russian. So, I was right about his accent sounding like Russian. And there is another small world connection. My Byler ancestry is Swiss, but before 1200 AD my ancestors lived in the mountains farther east, in the Balkans. According to DNA testing, I am four percent Balkan, and I carry a rare genetic marker which is found most commonly in the former Soviet Georgia and nearby Ossetia. Could Alexander Kartveli be a very distant cousin?

CHAPTER EIGHTEEN

The Fatal Crash of a C-130 on Take-Off at Bitburg Air Base

Very near the end of my three-year tour at Bitburg, I eye-witnessed another fatal crash on take-off. This time it was a C-130 cargo aircraft and the entire crew was killed instantly.

It was a Sunday morning in April, 1966. The C-130 had just delivered the 17th Air Force Band for a performance at Bitburg Air Base. The C-130 crew, for whatever reason, decided to practice a "max take-off" —which is to say that they would use minimum runway length and get up very quickly into the air—as one might do in a combat zone to avoid enemy gunfire. One may speculate about why that crew would do so, since Bitburg had a normal runway length with no abnormal obstructions at the end of the runway. There was no war on in Europe.

However, I later learned that the men on that crew were all Ohio Air National Guardsmen who were flying the "Atlantic Route" because the Regular Air Force crews had been diverted to flying the C-133 aircraft on the "Pacific Route" to ferry men to and from Vietnam. Perhaps those Air National Guardsmen were prepping in case they were sent to Vietnam.

On Sunday morning at Bitburg, I was, as usual, in the base chapel near the runway for Sunday School and Church service. We had just finished Sunday School in the basement of the chapel. Some of us were standing around, shooting the breeze, before heading upstairs for the morning worship service. Some of us were weapons loaders who knew each other.

Suddenly, one fellow weapons loader started jumping up and down, looking out the basement windows set high in the concrete wall, shouting excitedly, "He's going to crash! He's going to crash!"

I looked up at the window where he was looking, just in time to see the window fill with the roiling red and black of aviation fuel burning as an aircraft exploded on impact and was reduced to burning rubble. The flames looked like an exploding napalm tank.

We all rushed up the stairs and out of the chapel to see what had happened. A C-130 had crashed on take-off. The Crash Rescue crews never deployed. There was nothing to save, so they just sat in their Le France fire trucks and watched the crashed remains of a thoroughly destroyed C-130 cargo plane and a full load of fuel burn to the ground. The only thing left after the fire that resembled an aircraft was the vertical stabilizer, but even that displayed a rudder which had broken at the top hinge with the rudder tilting aft. The horizontal stabilizer was mostly intact, but that could not be seen from our ground-level vantage point.

The crew consisted of a pilot, co-pilot, navigator and the load crew. The passengers, the 17th Air Force Band, had all safely departed the flight line. Rumors afterwards had it that those bandsmen headed straight for the officer's club on base to down any number of drinks of alcohol—to quiet their jangled nerves. A close call indeed.

CHAPTER NINETEEN

Slippery Rock State College

It was a matter of good timing (another coincidence) that the Vietnam era G.I. Educational "Bill of Rights" passed right after I exited the Air Force. I had saved enough from my meager G.I. paychecks to cover the first year of college at Slippery Rock State College. I exited the Air Force in April of 1966. In June, the Vietnam era G.I. Bill was passed. Keep in mind that the educational benefit of the G.I. Bill did not pay up front. A veteran was paid in monthly installments only *after* the fact. I had to pay tuition *up front*. I needed the money I had saved plus the money I could make during a summer job.

Finding a job for the summer was tough. Universal Rundle where my father worked, made "sanitary ware" (bathroom fixtures: toilet bowls, lavatories, etc.) would not have hired me if they knew that I was only going to work there for the summer and would become a full-time college student that fall. (They had a policy against such hiring.) But I was able to conceal my intentions because I arrived home in April, not in June—and I was twenty-three by June of 1966—not eighteen.

Universal Rundle paid me the princely sum of $2.30 and one-half cent per hour—working a 40-hour week. That was a negotiated union contract. (I've always wondered about that extra half-cent per hour being negotiated.)

One day I had to go down into the sewer that exited the plant. I climbed down a ladder into the tunnel and scooped up buckets of the sewage from a small collecting pit, which was then pulled up by a rope. Some of the sewer slop dripped down on me. (Somehow, some of that seemed intentional.) I later learned that the storm sewer water and the sanitary sewage were co-mingled. Further, that raw sewage drained directly into the Shenango River. (Yes, that was already illegal even back then, and don't bother calling the EPA: Universal Rundle is just one more of the many companies that have failed and were shuttered permanently in the New Castle area.)

When the summer shut-down for two weeks came, the Rundle was surprised that I did not want to work clean up through that shut-down time. Everyone else among the common laborers needed to work to help make up for their meager wages. Somehow, I have never wanted to hear about my "white privilege." When a number of men of color were hired in response to criticism by the local NAACP, none of them stayed. They quit to the last man.

Each summer that I worked in New Castle, my wage per hour dropped: from $2.30-1/2 per hour to $1.90 the next summer, and so on. I drove truck in a quarry, stock-piling gravel with a dual-tandem dump truck. I would park under the chute of the separator unit, get out of the cab of the truck, and pull a large metal pole to open the chute to pour more crushed and screened limestone into the dump truck. That was actually a dangerous job. For instance, I was not issued a hard hat, even though one could be struck by falling stones, some of which were big enough to injure a person.

One time, lightning struck the steel structure of the gravel crusher. Had I had my hands on the steel pole used to open the chute to allow gravel to fall into the dump truck, I would have been killed by lightning. (That lightning strike shot through the whole crusher/separator system, frying many electrical components.)

At the end of the summer where I worked in the limestone quarry, I tried to get a withdrawal letter from the union so I would not have to pay an initiation fee again to the union if I came back to the same job the next summer.

I went to the union headquarters. In the office stood a tall black man—with a tell-tale bulge on his chest. The heavy-set man sitting behind the desk wore a shark's skin suit. His dark hair was slicked back, ending in curly hair. I thought I had accidentally walked into a movie set. I read the black guy as the armed "soldier." The guy behind the desk was too obviously the mafioso in charge of the union.

I was in my work togs: blue jeans and a cotton T-shirt. I was clearly unarmed—and not a threat--so I played it straight. I nodded a friendly acknowledgement to the soldier, then asked, politely, for a withdrawal letter from the union. The response was that I should pay my union dues through the winter, then come back the next summer and apply for work again. I said that would not be possible—that I would be a full-time college student and could not afford the union dues, then thanked them politely and excused myself. That was a "near thing." From my dad's twin brother came confirmation of my suspicions. I had indeed come face-to-face with the Mafia.

The next summer's job was in a meat-processing plant. Among other things, I had to clean out the bone barrels after they came back from the fertilizer plant and had had a chance to ripen in the sun for days. It was an odor that would have "gagged a maggot off a gut wagon"—as my father might have put it. Again, where was my "white privilege" in that?

In September of 1966 I entered Slippery Rock State College. I chose Slippery Rock because it was the closest state-owned college to my parents' home in New Castle—about 20 miles away. I intended never to commute because I wanted to experience campus life, and wanted to get on my own as much as possible. However, I needed to be close to where my parents lived so I would not have to travel excessive distance during college breaks and summer vacations. (The Vietnam-era G.I. Bill did not include a housing allowance such as the WWII version had.) Also, keep in mind that I was 23 and had been denied any reasonable opportunity for female companionship for four years. I needed a college degree, but I also needed a social life.

Going through registration on campus, I realized there was a problem. I was part of the first class of veterans coming to campus since the Korean War or WWII. A young female student who was working behind the table interrupted me with an annoyed, "You forgot to list your draft status on this form," handing it back to me. I replied, pointing to the bottom of the form, "I checked the 'veteran' box." Apparently, none of these young college students had seen a veteran before.

In truth, I still looked young enough to be an 18-year-old freshman, but I had been through a year at Penn State and a four-year enlistment in the Air Force. I was 23.

I took to wearing my polished combat boots and my Air Force fatigue jacket. (Hint. Hint.) One day I wore my olive drab Air Force fatigue shirt to class with my name tag and squadron designation from Bitburg. The professor approached me and asked, "Do you have a relative who is serving in the military?"

I replied, "No. This is mine. I am a veteran." He seemed more than a bit surprised.

I hit the ground running, academically. This time, I had to make it. The first semester, I made the dean's list, and I kept that streak alive, eventually graduating with honors (cum laude).

My senior year I was voted the "outstanding senior English major" by the English Honorary Society at Slippery Rock and was awarded a plaque stating such.

After I graduated, I went straight on into graduate school at Slippery Rock, graduating first in my class with a Master of Education degree in English.

I picked up a minimum-wage job on campus because I was one of the few students on campus with a camera. (Digital cameras and smart phones with a built-in camera were still future tense.) I had a Japanese-made 35 mm "Asahi" camera, and a twin-lens camera that took 120 film in square-format.) I worked for the campus newspaper, the yearbook, and ultimately, for the central administration and the sports information director—for minimum wage.

I soon parked the twin-lens camera and learned the timing for taking pictures of the various sports events with the 35mm camera. Football timing was different from basketball was different from baseball was different from track, etc. Then I took on the filming of football games with a 16mm movie camera. During the summers, I took wedding photos. Over the next several summers I photographed about 20 weddings.

Academics at Slippery Rock

Slippery Rock is often the subject of a bad joke because it has a funny name. Slippery Rock is named after the town of Slippery Rock—where it is located. It was originally founded as a two-year "Normal School" for teacher training. (Which, of course, also begs the question: "Is there also an Abnormal School?")

There are a number of state-owned colleges in Pennsylvania with similar stories—and some improbable names which also derive from the names of the small towns where they are located. These include Edinboro, California, West Chester, and Indiana—of Pennsylvania. That made for some interesting half-time football scores for broadcast on national radio. For instance: "Slippery Rock 21—Indiana 14." Many people around the country thought "Slippery Rock" was imaginary/made up.

I would wear a Slippery Rock T-shirt or sweatshirt when I traveled around the country. Sometimes someone would approach me and ask, "Is there really a Slippery Rock?" I would respond, "There had better be: I have two degrees from there!"

The popular myth is that you have to go to Harvard, Yale, etc. to find good professors to get a good education. I had some world-class professors at Slippery Rock. For instance, two of my writing teachers are world-renowned today: Brendan Galvin and Ken Smith. Today, Brendan Galvin, PhD in writing, is regarded as the finest American poet still alive and writing today. Ken Smith passed away (prematurely) in 2002, but he is regarded as the finest English poet in the latter half of the twentieth century. Ken Smith and Jon Silkin (another highly regarded English-born poet) co-founded "Blood Ax Books."

I had a linguistics professor, Dr. George Waldo, who earned his PhD degree from the University of London and who was one of the professors who helped found the university in Malaysia. He knew all of the notable lexicographers in the dictionary publishing world. My general science professor, Dr. Ward (PhD), discovered the oldest glacial till found in Pennsylvania—and named it "Slippery Rock Till." One of my English teachers came from India. A brilliant man, he had published over 200 journal articles in the area of comparative literature: American and Indian (as in "dot" not "feather"). Some of his papers won literary prizes.

The theater department at Slippery Rock staged the first non-professional production of *On a Clear Day*. They also had an offer to produce *Hair* but the administration at Slippery Rock knew that would never do in such a conservative small town as Slippery Rock. (The campus priest was the brother of Gerome Ragni, one of the co-authors of *Hair.*)

Several theater grads from Slippery Rock went to New York City and had professional careers there as actors. One grad became a professional dancer, also in New York City. Another actor and singer, from Pittsburgh, became "Neighbor Aber" on Mr. Rogers. (His real name is "Chuck Aber.") As campus photographer, I photographed Chuck Aber for the campus newspaper multiple times.) One female grad later won a "best supporting actress" award in a movie about the Holocaust. One theater-major who became a public-school teacher memorized, uncut, the lines of "Richard" in Shakespeare's "Richard the Third." (He never needed a prompter or missed a line.)

The chair of the English Department, Dr. Daniel Marder, had a vision for a creative writing degree at the master's level for Slippery Rock—instead of a cookie-cutter English master's degree which mimicked all the other state colleges in Pennsylvania. To that end, he recruited the likes of poets Ken Smith and Brendan Galvin, plus other published writers of fiction and poetry. He hired Isaac Singer's translator to teach on campus, and invited Singer himself, John Ciardi, and several other nationally-known poets to speak and read at Slippery Rock while I was a student on campus.

Dr. Marder, the chair of the English department, recruited me into that proposed creative writing program with the promise that I could write a "creative thesis" rather than doing a traditional research project and master's thesis. Sadly, the powers-that-be in the state system of higher education in Pennsylvania did not approve that proposal. The linguist who mentored me through my master's degree thesis was fired, and the poets and other writers were forced to seek employment elsewhere. Dr. Marder, the chairman of the English Department, moved on, and I was encouraged to do a conventional master's degree research study and paper.

I wrote a short story for the campus literary magazine, *Ginger Hill,* at Slippery Rock which won "first prize for fiction" in the Spring 1969 issue (Volume V, Number 2). Brendan Galvin (one of the judges) told me I was a writer; the first person to do so. That short story was closely based on the actual fatal crash of an F-105 at Bitburg Air Base during my first Operational Readiness Inspection at Bitburg. (See my renamed and lightly edited script of that short story at the end of this memoir.)

Ken Smith wanted to recruit me as a photographer to accompany him on a tour of the US, but he never was able to find someone to bankroll such an enterprise. Ken Smith was fond of such travels to gather new material for his poetry. After leaving Slippery Rock, he taught at two colleges in New England and then went to dangerous places in the world (such as Eastern Europe when it first opened up).

One such trip (to Cuba) ultimately caused his death. First, he was stopped and harassed by police in Cuba who thought he must be some sort of spy—or up to no good. He finally convinced them that he was simply a poet and a British citizen who wanted to see Cuba to collect new material for his poetry.

He contracted Legionnaire's Disease while in Cuba, but made it back to a hospital in England. There, he developed pneumonia and died—a sad ending for a great poet, and a really great person.

I ended up doing a linguistics research paper at Slippery Rock entitled: "An Historical View of the Stress Patterns of Two-Syllable English Verbs from 1755 to the Present." (The longer the title, the smaller and less-significant the subject.)

Let me say a few words about my linguistics professor, Dr. George Waldo. When he was teaching in Malaysia, helping to found the university there, a Chinese female student asked him a question: "What is the system in English for the placement of stress on syllables?" Dr. Waldo thought for a moment, and then said in reply, "That book hasn't been written yet." (I would note that that book to this day apparently has not been written.) It is, after all, a rather important subject, for if one places the emphasis on the wrong syllable in English, no one will understand what you are saying. (Try, for instance, placing the stress on the *second* syllable of "emphasis" or "syllable" instead of on the first syllable.) Interesting? There are other quirks in English such as words like "record" which change from noun to verb depending on which syllable is stressed.

Some words get different stress placement depending upon which side of the Atlantic Ocean one resides on. (There is an old joke about "Great Britain and America—two nations separated by a common language.")

Dr. Waldo encouraged me to apply to several doctoral degree programs in linguistics, and to expand my study of two-syllable verbs into a doctoral dissertation. I did apply, and Ohio State accepted me into their doctoral degree program, but several things bothered me about continuing on into such a doctoral degree program. First, the various universities did not seem to be sure to which department the linguistic study of English belonged (was it the English Department or the Foreign Language Department?) Also, the fact that the writing talent at Slippery Rock was forced to move on bothered me. Also, the fact that my linguistics professor was fired (was there really a just cause?) bothered me. Incidentally, he never again taught on the college level. (I think they must have black-balled him.) Why? He was a perfect gentleman and a brilliant scholar.

There was also another issue: I was already 28. I figured I was going to become a "professional student" (for eight or ten years), then collect an unmarketable degree and graduate into unemployment in my thirties. Though I did have that acceptance into a doctoral degree program at Ohio State, I took my teaching certificate in Secondary English and my Master of Education degree in English and went searching for a high school teaching job. It was high time I got on with my life and a teaching career.

The job market was very tight. I was a veteran who went to college with a large group of young men who had enrolled in college as a legitimate dodge from the draft (at least for four years). The War in Vietnam raged during those years. Most young men in Western Pennsylvania couldn't afford an upper tier college or university. The more affordable colleges were the state-owned "teacher preparation" colleges like Slippery Rock, etc. So, that is where those young, draft-eligible men enrolled in college—in droves.

We all graduated at the same time when older teachers were hanging on to their jobs for dear life, and there were too many young men (and women) competing for the very few openings in the school classrooms.

Again, I had graduated with honors for my undergrad degree. I graduated first in my class for my Master of Education degree in English. I had my temporary teaching certificate. I had fifty job applications out all over Pennsylvania, plus a few in Maryland and up in New York State. I received only four interview invitations. One was clear up in southern New York State and another in the extreme northern tier of Pennsylvania. One district in Southeastern Pennsylvania, where I had interviewed, wanted to offer me a job, but misplaced my application—until it was too late: (I had already signed a contract in Delaware County, in south eastern Pennsylvania.) The two interviews in Southeastern Pennsylvania required two separate 300+ mile drives across Pennsylvania. Also, I had to hurriedly find an apartment to rent for the coming school year. The fact that the second school offered me a job contract right after the interview was completed should have been a warning sign that they were a little too eager. Red flag. But I needed a job, so I signed.

CHAPTER TWENTY

My First Teaching Job

That district where I did sign a contract lied to me and my fiancé in that the superintendent handed my bride-to-be an application form—knowing that they would never hire her. They had an unwritten rule against hiring ANY relatives, not even cousins, let alone husbands and wives. (I never discovered the story behind that unwritten rule.) Also, I later learned that the school district was paying for three superintendents: the one who was filling the job—plus two they had fired. That was certainly another red flag.

The school was a nightmare: pupil and teacher absenteeism were quite high, another sure sign of a school district in trouble. (By the way, I never missed even one of the 180-day teaching schedule that year.) The high school was dealing with twice the number of students the building was designed to hold. Discipline problems were rampant. Twice I was threatened with bodily harm by two different male students—in front of two different assistant principals (one of whom was the school disciplinarian.) Neither of the two assistant principals did anything in response to those terroristic threats. The first time that happened I was too stunned to say or do anything. I was literally speechless. The second time that happened, I looked at the assistant principal, then at the student and said, "In the first place, you do NOT threaten a teacher. In the second place, if you start on me, I will wipe the floor with you!" (I meant every word of it). Apparently, once again, my age and background were underestimated. I was 28, had been through a four-year enlistment in the Air Force (where I ended up with the rank of "sergeant E-4" as a weapons loader—some of the toughest, strongest men on the air base). I could carry 100 pounds of 20mm ammo in each hand and walk across the flight line. And some of my weapons loading buddies were twice as strong as I was. I saw one of them pick up the rear end of a Mercedes sedan by himself and hold it off the ground.) I had a year at Penn State, plus five years at Slippery Rock (four years undergrad plus a master's degree).

When I checked the evaluation records of my students, I was shocked. Because of over-crowding in the school, I had an over-lap schedule between the morning and afternoon sessions. The three morning classes were labeled "low academics"—which meant that the students were capable, but not interested in school/ did not want to do the academic work. The two afternoon classes were eleventh-grade boys whose average reading level was third grade! Meanwhile, my teaching materials were written on the college level. (I know because I checked those teaching materials against a reading-level scale.) It was hopeless. Some of those male students were totally illiterate. By the way, none of them belonged to any of the cliché underprivileged racial or ethnic groups. (They were all "lily white" and working class.) Only one young man was on grade level for reading competence and he spent his time acting as dumb as the rest of the class, including intentionally slurring his words and reading very slowly. (He did not want to stand out as being different from his classmates.)

The only bright spot in that school year was my landing the job of "head varsity rifle coach." The previous coach, Frank Geno, had just retired. He had the incredible (unheard of) record as a coach of one-hundred-and-eleven straight wins in dual-meet inter-scholastic competition. He hadn't lost a single match in dual-meet competition in decades, plus he had a string of district championships and several state championships along the way.

The story around town was that the rifle team would "never win another match with this new coach." However, we went "clean" for the season, winning all our matches, including the "District One" title. (Who ever heard of a freshman coach going undefeated for the year—in any sport?) We missed the regional championship title by ONE very slim point: their winning shot just barely touched the ten-ring, while our losing shot was just the thinnest pencil line outside the ten-ring. The team which beat us went on to win the "Pennsylvania Scholastic Prone State Title" by a good margin. We came so close to getting a chance to win that state title.

So, unsatisfied, I took my students, divided into three teams based on grade level (10th, 11th, and 12th), and entered them in the (completely separate/outside the public school system) "Pennsylvania Rifle and Pistol Association State Prone Championship." We won that title—including the fact that one of my girl shooters had the highest score in the entire state that year—for both male and female shooters.

I have always been proud of the fact that target shooting was one sport where there was no discrimination by sex in competition. Men and women in the US compete as co-equals, on the same ranges, with the same equipment, on the same targets. The "girls" often beat the "boys." (I would note that eight of my top ten shooters on that championship team that year were young women.) Only the Olympic and "International" target-shooting teams separate men and women in target-shooting competitions. (One may thank Russia and Third-World countries for that discrimination.) Such sexual discrimination has never been part of American target shooting competition.

The next academic year, the head coaches of the other varsity sports got their heads together and reduced "riflery," as they called "target shooting," to an "activity"—not an "inter-scholastic sport." (They didn't want boys and girls competing equally.) "Title Nine" was just coming into force in Pennsylvania. I personally appealed to the state of Pennsylvania governing body but got absolutely nowhere.

How can it be that girls shoot as well as or better than boys? First of all, girls will listen to a coach. Boys generally think that they already know it all—especially when it comes to handling a rifle. Secondly, girls shoot the same under match pressure as they do in practice. Guys get hyped up under match pressure with resultant higher pulse rate and blood pressure—both of which hinder performance in target shooting.

One of the most unbelievable records of all time in target shooting was shot by Mary Stidworthy when she was a young single woman. She put 205 straight shots in the "X" ring of the target (dead center) with a .22 caliber rifle at 50 yards. (That is outdoors, exposed to wind changes, temperature changes, and changes in the sun light and mirage.) Every twenty shots, the shooter has to walk "down range" and change the target. That is an incredible feat of marksmanship and endurance after a long day shooting 160 shots for record in four separate matches at various distances from 50 yards to 100 yards. And she did this with a "Model 37" Remington, so named because it was designed in 1937. Sure, she had a custom barrel, a custom trigger, and a custom fitted stock, but the "lock time" on more modern target rifles is a fraction of that from 1937. Quicker lock time is a definite advantage in target shooting.

By the way, as a junior shooter, I earned the National Rifle Association's highest level of qualification, "Distinguished Rifleman." At Penn State I had the third highest average on the Air Force R.O.T.C. rifle team as a freshman, and we won the Inter-service match against the Army and Navy R.O.T.C. teams at Penn State. In the U.S. Air Force, I qualified for the Air Force "Expert" ribbon for "rifle"—the only level of marksmanship recognition which the U.S. Air Force offers—unlike the other military services which offer medals for several levels of shooting proficiency. As an adult, the NRA awarded me the rating of "Lifetime Master" in small-bore, three position, outdoors and I had "master" ranking in several other categories of .22 caliber target shooting, both indoors and outdoors. I also have NRA certifications as both a firearms safety instructor and as a marksmanship training instructor. Those last two certificates played a part in my being accepted for the position of head rifle coach. I would have liked to have had another crack at coaching that same high school's rifle team, but I knew it was time to move on to another teaching job.

By the way, it was my student equipment manager (who didn't quite make the rifle team) who wrote to me the following year to tell me that my teaching replacement lasted exactly one day. At the end of that day, he placed his class lists and grade book on the principal's desk, announced, "I quit" and never returned.

Again, coincidence intervened. I had a sister-in-law who was working in the Boyertown Area School District in Southeastern, Pennsylvania as a math teacher. She notified my wife and me that her district had openings in both their English and Foreign Language departments. My wife had an M.A. in Spanish and an undergrad degree in French—a versatile dual-threat in most school districts. She had graduated summa cum laude as an undergrad, had a teaching certificate, and had also taught as a graduate student at the University of Illinois, Champagne Urbana Campus, as she was earning her M.A. degree. I had a master's degree in English and a teaching certificate. We were both hired.

CHAPTER TWENTY-ONE

Boyertown Area Senior High School
Small World, Again

Boyertown, as it turned out, is the birthplace of General Carl A. Spaatz (pronounced like "spots": it is, after all, a Germanic name), the last Chief of Staff of the Army Air Corps and the first Chief of Staff of the newly independent US Air Force. (Isn't it peculiar how the Air Force keeps cropping up in my story?) General Spaatz began his service in WWI, where he earned the Army's *Distinguished Service Cross* (second only to the *Medal of Honor* as an award for combat valor—and exactly equivalent to the Navy Cross and the Air force Cross) for taking down three German Fokker fighter planes on one mission. (He had to step down from his command position to become an ordinary "pursuit pilot" in order to shoot down those German fighters.)

In 1929, Spaatz was on the receiving end of the fuel hose (and commander of the operation) in a record-setting endurance flight to prove the efficacy of mid-air refueling. That earned him the coveted Collier Trophy and a *Distinguished Flying Cross.* During WWII, as a General he was the one who convinced General Eisenhower that strategic air power would win the war, rather than using aircraft for tactical purposes. Spaatz was right, and his philosophy of air power brought WWII to an end much sooner than the alternative would have—and saved countless lives in the process—on both sides.

Before General Spaatz retired, shortly after the newly separate US Air Force was created, he set up the "major commands" in the Air Force: Strategic Air Command, Tactical Air Command, Air Defense Command, Air Training Command, and Military Air Transport Service (SAC, TAC, ADC, ATC, and MATS). [Mats later became MAC—Military Airlift Command]. That command structure remained in place for a half century, until after Desert Storm. The changes that were made at that time let our guard down in the area of air defense over the Continental US and 9-11 happened as a consequence. (That would never have happened under General Spaatz's watch.)

When those two planes crashed into the World Trade Center buildings, we did not have one single fighter plane in the U.S. loaded with live ammo in the guns, nor did we have a single aircraft loaded with a live interceptor missile. (Apparently, the civilians in control of our military did not trust our pilots with live munitions.) Two F-15 pilots (one male, one female) who were sent up from "First Wing" in Washington, D.C. to intercept the airliner which was heading for our US Capitol were on a suicide mission. They had no missiles nor were their Gatling guns loaded with ammo. One of those two pilots was to crash into the cockpit of the passenger plane. The other pilot was to crash into the tail of the passenger plane. Their mission was to keep that passenger plane from getting to Washington, D.C. Again, that would never have happened under the watch of General Spaatz. His Air Defense Command would have been armed and ready. After 9-11, Air National Guard F-16s were loaded with live ammo and put on alert status for immediate launch if there were another passenger plane hi-jacked and headed for a target in the US.

Boyertown High School

Boyertown Area High School was a big improvement over my first year of teaching. A growing school district population in the Boyertown area was pressing the limits of classroom capacity. My classroom typically had 40 pupils per class session—because that was how many chairs the classroom could hold (five rows of eight chairs or 5 X 8 = 40). Five class sessions per day times forty = 200 students which I was responsible to teach and test. (Have you ever tried to grade 200 composition or tests?) And still there are those who claim that "teaching is a whipped cream delight of a job." As one of my fellow teachers, a Marine helicopter pilot once said, "If you think that teaching is such a good racket, get in it yourself." (As I write these lines, schools all over the country are in need of teachers.) If you have a college degree, step up to the task. I am too old. (I am now eighty.)

One time a man who worked a typical factory job 40 hours per week told me, "Ah, you teachers go home, put your feet up, and watch TV just like everybody else." Really? My first year at Boyertown, I had five different preparations to get ready each night for the next day's lessons. In fact, I spent four hours or more each night prepping lesson plans for the following day or grading papers or tests. Each quarter, I had five new "quarter courses" to prep and teach.

To make matters more challenging, just before Thanksgiving, my wife and I had a fire which started in the furnace room under our first-floor apartment. We essentially lost everything, including my testing materials, books, lesson plans, etc. We had to start all over again. Anything not outright burned was ruined by smoke, heat, and water from the fire hoses—including our TV and furnishings. Our brand-new sofa bed was also ruined, as well as a hand-made quilt which was a wedding gift from my wife's grandmother. It could have been worse. Later, I learned that the coach of another high school rifle team I had competed against the previous year had also had an apartment fire. He lost his wife, and sustained burns and permanent scars to his hands and face in a vain attempt at saving his wife's life. She had escaped the fire but rushed back into the burning apartment, apparently to try to save some valuables. The smoke overcame her, and he was not able to find her in the dense smoke. He just barely got back out alive and spent months in recovery.

Then we had to fight the insurance company over the meager $4,500 insurance we had as renters. (They claimed that we over-estimated our losses, but our losses were easily double that insured amount, which was the maximum for renters.) We spent the weekend washing the few underclothes we had in drawers and seeking a new apartment elsewhere. We had to immediately buy a new mattress. My shoes, my good German boots, my expensive German overcoat, and all my clothing on hangers (pants and shirts, as well as coats and jackets and suits) were burned. My wife's polyester dresses (in a separate closet from my clothing) were turned into hard plastic by the heat of the fire. I lost two antique shotguns, plus my modern hunting shotgun. The floor of our closet burned through and the remains of that shotgun were found in the basement—with the wooden stock completely burned away. My deer rifle had to be completely refinished, both the metal and the wood. The heat of the fire completely collapsed the magazine feed spring. My prized Winchester Model 52 target rifle was protected by a wooden case, but was peppered with rust spots from condensation caused by the heat of the fire. My new leather shooting jacket smelled heavily of smoke, as did my shooting mat.

Long after the fire, I kept remembering things that we had in that apartment which I had forgotten to list for the insurance company.

A neighbor in the next apartment building loaned me a pair of pants and a shirt to wear. Fortunately, we were the same size. The Red Cross gave us a check for $100—for food and clothing. I bought two pairs of dress pants and two shirts to get me through the next week of school. (The only pair of pants I got out of the apartment with were my blue jeans I had put on before we fled the smoke and fire. (One could not wear jeans as a teacher.)

Twice during my teaching career in Boyertown, I had to take a long-term leave of absence for medical reasons. (Both were surgeries for torn rotator cuffs). Both times, my substitute lasted one day—and quit.

At that other school where I taught my first year of teaching, my replacement at the start of the next school year also lasted one day, as noted before. (He received the same schedule as I had the previous year, the same schedule that sent the young female teacher packing that I replaced.) You would think school districts would get the message and change something. But, no. Note also that the principal of that high school in Delaware County Pennsylvania had promised me "Next year will be better." However, he was lying to me: my replacement received the same schedule as I had. That young man threw in the towel after one day.

CHAPTER TWENTY-TWO

Civil Air Patrol

I joined the General Carl A. Spaatz Squadron of Civil Air Patrol in Boyertown, Pennsylvania. That squadron was the first in the nation to be named after a person. The squadron commander was Colonel Elizabeth Magners. Her husband had served as a pilot in WWII and Korea. He had been the Civil Air Patrol squadron commander. After his passing, she took over the command of the squadron. Her service with Civil Air Patrol had begun during WWII when she was a "cadet," and would continue for over 50 years. Because of her leadership, I eventually wore the rank of "captain" on my Air Force uniform as a Civil Air Patrol officer. In all, I served Civil Air Patrol for over twenty years both as an "aerospace education instructor" and as a "certified firearms safety instructor and marksmanship training instructor."

Her son Richard was a student in one of my English classes, and he invited me to join Civil Air Patrol. Richard was an outstanding student, one of many I had in my classes back then. Richard would go on to graduate from the Naval Academy and had a good career going in the Navy until his wife decided she did not want to be left in port in San Diego while he was at sea as captain of a USNS ship (commanded by a Naval officer but manned by civilians.) Further complicating her life was the fact that though she had a teaching certificate from Pennsylvania, she could not get a job teaching in California without being proficient in Spanish. She finally forced him to resign his commission and they returned to Pennsylvania to seek employment. Sadly, she ultimately left him for another man and divorced Richard. He was half way to a military retirement when he resigned his commission. Sad indeed.

Besides Richard Magners, I had four other young men over my 26 years of teaching in Boyertown's schools who went to the U.S. Military Academies: one more to the Naval Academy, one to the Coast Guard Academy, one to West Point, and one to the Air Force Academy. Still, some claim that Boyertown High graduates "do not get good scholarship awards." Pardon me, but getting an appointment to any of the US military academies is a million-dollar scholarship: a four-year degree, an Academy ring on the finger which will get one in the door with many firms and companies, plus free medical care (now that is "Tri-Care"), good pay, and a good retirement pension for life.

These last two young men where in my son Chad's graduating class. The one who was president of his high school graduating class went to West Point and became the secretary of his class at West Point. The one who was vice president of his class in Boyertown became the president of his class at the Air Force Academy. I was very impressed. I thought such positions of honor were reserved for the sons of generals and the like. Neither one of those young men had a father who was an officer in any branch of the US military and neither of them had any political connections. They got their appointments to their respective military academy on their own merits and they were elected by their peers on their own merits. That is impressive indeed.

One story of coincidence about the young man who went to the Air Force Academy

I loaned John, the young man who went to the Air Force Academy, a signed copy of BGen Robbie Risner's autobiography, *The Passing of the Night*. I told him it was required reading at the Air Force Academy, and that he should read it before arriving there. BGen Risner was a decorated three-war veteran. Like so many, he spent long years as a POW in Hanoi. He was also an "ace" and a master of "dog fighting." His last command was as an "air boss" of three wings of F-111s. (James Kasler was deputy commander of one of those three wings.) Small world, again.

Brigadier General Robinson Risner was one of the many people I interviewed for Colonel Kasler's biography. I spent a long afternoon interviewing him when he lived just south of Austin, Texas. Like Colonel Kasler, he flew the F-105. Risner was shot down twice. The second time, he was captured, and spent long years as a POW in Hanoi.

Ross Perot paid to have a life-size statue of General Risner made. In one more coincidence, that statue was dedicated at the Air Force Academy that fall when my former student from Boyertown arrived there in time for that ceremony.

He wrote to me, asking; "Mr. Byler, What did you *really* do in the Air Force?" I had also instructed him (and the other young man who went to West Point that year) in target shooting and in marching drill. Neither of those two young men had any prior instruction in marching or in target shooting.

I wrote back to him: "No, John. I was just an ordinary enlisted man who served as a 'weapons mechanic' and 'weapons loader' on the F-105 *Thunderchief.*"

After sixteen years teaching in Boyertown Senior High School, I needed a change, and therefore requested a transfer to Boyertown Junior High West. (People thought I was crazy.) Eighth graders have been described as "hormones on feet." Many of the boys seemed very immature. Some of the girls, on the other hand, were my height (eyeball-to-eyeball with me at 5'9"). They looked like grown, fully-developed women, but they certainly were not. Immaturity cropped up in unfortunate ways.

The principal at Junior High West decided that he would introduce me to the junior high students by having me play the grand piano (a piano, coincidentally, which had been played by Liberace—and was signed by him) in the auditorium as the students marched in for an assembly. I played "Misty." As it happened, there was one young lady/eighth grade student who just happened to be assigned to one of my English classes, whose favorite song was "Misty." (Had I known that, I would have selected another tune.)

"Misty" is also one of my favorite tunes. It has been a pop music classic for many years. It was also Colonel Bud Day's favorite song. Bud Day founded and commanded the "Misty FACs" (forward air controllers) in Vietnam. They flew F-100s on the Ho Chi Minh Trail. Bud was eventually shot down and captured. He would spend long years in prison in North Vietnam under severe deprivation and brutal torture. He wrote *Duty, Honor, Country* about his experiences. (No publisher wanted to touch it, labeling it "too red, white, and blue" so he self-published it.) Never mind that he earned the Medal of Honor for his combat valor and his endurance under torture in Hanoi, plus an award of the Air Force

Cross, etc. Later, Robert Coram wrote *American Patriot* about Bud Day's life which was published by Back Bay Books/Little, Brown, and Company, 2007. As a lawyer, Day argued before the Supreme Court for medical rights promised to veterans of WWII—while wearing his Medal of Honor around his neck. And, yes, he also served in combat in WWII and Korea.

See: *American Patriot* by Robert Coram: Back Bay Books, Little Brown, and Company, New York, 2007

Back to that junior high girl student

Her parents had divorced. Had I known that I would have avoided her as completely as possible. I eventually learned that young women who lose their fathers, either through death or divorce, tend to find an older man to replace their missing father. This eighth-grade girl/young woman had picked me—as her fantasy "father." (Oh no.) Not only is it unethical to get "involved" with an under-aged student, but it is just plain illegal: the kind of illegal which will not only get a teacher fired, but will also cost a teacher his teaching license, his pension, and his freedom. In Pennsylvania (and most other states), "seduction by teacher" is classified as statutory rape. No ifs, no ands, no buts. Further, young women often live in a fantasy world, and they share that fantasy with their female friends—who then tell their boyfriends. I became public enemy number one among the young males—who were typically beginning to "feel their oats" and perceived me as a competitor.

While I was teaching at the high school, I had a parallel experience. One girl with an over-active imagination fantasized that I was her "boyfriend." She was vocalizing that fantasy with her female friends when she was overheard by another male teacher. He took her aside and told her that if she continued to talk like that, I would be fired. She apparently got the message and stopped such talk. Ironically, that male teacher (who likely saved my teaching career) was later fired for having had "relationships" with several female students. How ironic.

CHAPTER TWENTY-THREE

Meeting BGen Robert L. Scott

In 1986, I met BGen Robert L. Scott, quite by accident, and John Frisbee wrote a "Valor Page" in *Air Force Magazine* 69, no. 11 (November 1986): 119. Those two events in tandem changed my life. Furthermore, if I had not been sent to Bitburg Air Base I would not have known that Robert L. Scott was the first commander of 36th Wing at Bitburg. If I hadn't met Brigadier General Scott, I would not have written to John Frisbee, the author of the "Valor Page" article in *Air Force Magazine* about Colonel James Kasler. If I hadn't written to Colonel Kasler through John Frisbee, I would not have learned where Colonel Kasler lived in Momence. Illinois. What a series of improbable coincidences.

In the summer of 1986, I was heading south from Pennsylvania to Savannah, Georgia to visit my sister Sally for the first time. She and her husband Bill spent their entire married lives in the Savannah area. (Brother-in-law Bill passed away from pancreatic cancer a few years ago). As we passed by the Warner Robins Air Museum, I spotted an F-105 sitting along the highway as a drawing-card for the museum. I said to my wife, "We have to stop there on our way home."

Of course, I had worked on the F-105, and the F-105s had all been retired from service in 1984. The F-105 quickly disappeared from the scene. Just a little over 100 examples out of the original 833 survived—mostly in museums or as "gate guards" at air bases, or at VFW posts, public parks, and the like. However, there were not yet published books about those survivors and where they were located. I was actively looking for them. Many F-105s were intentionally (and needlessly) destroyed, including some historically significant ones. I tried, fruitlessly, to get my local air museum to go after at least one F-105 parked at Aberdeen Proving Grounds. The local museum could have had the F-105 for the cost of transporting it to Southeast Pennsylvania. That was an ex-Bitburg plane which had also served in the Viet Nam war. Because of its Bitburg history, it carried *Mad German Express* lettered on its left intake while serving in the Vietnam War. Monogram produced two accurate

1/72 scale model versions of that relatively-rare twin-cockpit model: one as an "F" model as it appeared while serving at Bitburg: with silver paint finish and three tail stripes (red, blue, and yellow) of the three fighter squadrons at Bitburg (22nd, 23rd, and 53rd), plus the tail number 38304. The other Monogram model was the same F-105 converted to a "G" model "Wild Weasel" in camouflage paint as it served in Southeast Asia. That model carried the unit identifier "WW" on its tail along with its serial number. Included in the Monogram kit are two "Shrike" missiles and one "anti-radiation" missile for killing a SAM site.

Sadly, that distinguished combat veteran's jet engine was fired up at Aberdeen Ranges as a heat source, sitting on the ground, and a "Sidewinder" heat-seeking missile was fired into it, burning it to the ground. (Sob.) To what purpose? As though they didn't already know what happens to a jet fighter plane when one shoots a Sidewinder missile into it?

Aberdeen also had two rare twin-seat F-100-Fs—one of which could have been decorated (painted) as Bud Day's "Misty FAC" as he flew it in Vietnam. No such luck. Sad, again.

By the way, that air museum in Reading, PA still does not have a single fighter plane from the so-called "Century Series" (F-100, F-101, F-102, F-104, F-105, F-106). They do have an F-86 that went to the Spanish air force after it served at Bitburg Air Base. However, after Spain was done with it, they took off the wings, put one forklift in the nose intake and another in the exhaust. When they lifted it, the wingless fuselage rolled over and crushed the canopy. They virtually trashed it.

On the return trip from Savannah, we did stop at Warner Robins Air Museum to look at their two F-105s. That's when I discovered that this Air Base was where the famous Flying Tiger "ace" Robert L. Scott began his flying career during WWII. As I was reading about Scott's illustrious career in a show case, I heard two older gentlemen talking right behind me. Though no names were mentioned, from their conversation I could tell one of those two men was General Scott! I waited until they were done talking and then turned around. I walked up to a tall, slender, white-haired gentleman with a hawk-like face who was still standing there and asked, somewhat incredulously, "Excuse me, sir. Are you General Scott?" He replied, "Yes. I am." (I had seen photos of Robert L.

Scott, but those were generally of him as a younger man when he was a Flying Tiger in WWII.)

I was stunned. I didn't know that he was still alive, and I certainly was not looking for him. Keep in mind that Robert L. Scott was thought to be too old to be a pilot in WWII, and now, decades later, he was *really* old! This series of coincidences was too improbable to ignore.

When I make presentations in churches, I sometimes label my presentation, "No Burning Bush" (like Moses saw). Nor was there a bright light such as St. Paul experienced. There was no bolt of lightning, nor was there "a still, small voice." But I got the message. "Pay attention, Chuck. This is important."

I later traded letters with General Scott.

That November, the second half of the message came when John Frisbee published his "Valor page" in *Air Force Magazine.* I did what would have been unthinkable prior to meeting General Scott: I wrote to Colonel Kasler through John Frisbee, the author of the "Valor Page" at Air Force Magazine. A while later, Colonel Kasler sent me a letter in reply. Now, I knew his address/where he lived.

The next summer, on vacation from school teaching, I drove out to Colonel Kasler's golf club just outside Momence, Illinois. Over lunch in his club, he recounted the story of his first super-sonic dive in an F-86A in Korea. That was nearly the end of him. Once he accelerated to near the sound barrier in a dive, he lost the ability to pull out of the dive, but he remembered what he was told: use the trim tab to get out of a dive at or near the speed of sound.

However, after his landed, the F-86A was a total mess: wings bent and one flap hanging down. That fighter plane never flew again.

I knew then that Colonel Kasler had a story to tell and that I wanted to write it, but how to convince him that I had the ability to do the job? After all, he had told a number of people that story before, and no one had taken the initiative to pursue writing his biography.

CHAPTER TWENTY-FOUR

A Second Master's Degree – in Writing

In 1990 I was eligible to take a sabbatical leave from teaching at half pay. I had been thwarted in my intentions to do a master's degree in writing at Slippery Rock. Over the years as a teacher, I had collected any number of experiences. My first cousin had served in Vietnam as a Marine, then in Washington, D.C. as a street cop. I had vague notions that I could write a novel as my "major product" for a Master of Arts in Writing Degree. Maybe I could use that to prove to Colonel Kasler that I could produce a book-length work.

I was hearing, off and on, stories from guys who were combat veterans or cops or both. Something began to resonate, and I felt like writing about it. I began looking for a "low residency" writing program. (I didn't want to abandon my wife and two kids for a whole year.)

I even considered commuting to Penn State, but they seemed uninterested in me. When I told them I had a recommendation from Brendan Galvin, the professor on the other end of the line said, dismissively, "But he is a poet—not a biographer or novelist." I bit my tongue, for I felt like coming back with "When is the last time when you were a finalist for a National Book Award?" But I did not. I have learned that it is futile to argue with "experts." Just chalk it up as one more negative experience with Penn State University.

Finally, I found a program for adults up at Vermont College—which at that time belonged to Norwich University—the only "militia" school in the Northeast. I called that college to enquire about their writing program, but was connected to the wrong phone number—which turned out to be the right phone number for what I was seeking.

I attended regional meetings for that program—testing the waters. I found myself surrounded by many lesbian women and a few homosexual males. The women really seemed to have a large chip on their shoulders—no doubt based on bad experiences with heterosexual men. I dove in anyways. Perhaps I could prove them wrong about their prejudices about men. However, in one of the open discussion sessions, one of the women stated flatly that "Eighty-five percent of heterosexual men are rapists." No one objected to that sweeping statement. I was stunned. I knew that was wrong/incorrect, but I remained silent.

For years I wondered why the "85 percent" figure. Finally, one day it dawned on me. Those women believed that homosexuals numbered fifteen present of the population. Of course, in their logic, that 15 percent could not possibly be rapists. Oh my.

I needed a field faculty advisor with a terminal degree in my preferred area of study—since this was a "make-your-own-degree" program. I needed someone with a terminal degree in writing who had taught at the college level. Technically, there was not yet a PhD in writing at that time. (There is now.) However, I found David Chacko, a published novelist who had taught at the college level. That, again, is a story of peculiar coincidence.

I knew a fellow high school art teacher at Boyertown High named David Larson. He taught right down the hall from my classroom in the high school. David is a "water colorist" and had served as an infantry lieutenant in the Army. A draftee, he had done a tour in Vietnam. He was fond of going to a used bookstore in a local flea market. One day he found a novel called *Brick Alley*. It was set in a small town near Pittsburgh in Western Pennsylvania. David knew the territory because he was raised in that area. He knew that "Brick Alley" was a legendary "red-light district" in that town near where he was born and raised.

David Larson had heard my stories of tracking down people, so he set about trying to find the author of *Brick Alley*, one David Chacko. One thing led to another. David Larson tracked down David Chacko and I eventually made contact with him. I talked Chacko into being my mentor/ field faculty advisor for my Vermont College degree. In exchange, I became his technical advisor on firearms for his novels. David Chacko had been in the Air Force, but he did aerial mapping the old-fashioned way—before the advent of satellite mapping. He had no experience with firearms beyond Air Force basic training in marksmanship.

When Chacko wrote a fictional account of the killing of Hitler's second-in-command, Reinhard Heydrich, by a Czech commando squad, he was puzzled by why/how the Sten Gun failed to fire. So, I explained how a Sten Gun operates and why it likely failed to fire—and he dedicated the book to me as a thank you. (The title of that book is *Like a Man* by David Chacko.) I have both editions of that book, including the second edition which has a death mask of Heydrich on the cover.) It is a very good read, and the title has multiple meanings.

While he was living in Florida, Chacko wrote a novel set in Florida. He needed a name of a specific revolver that a woman might carry. I suggested a "Lady Smith," a small-frame revolver which Smith & Wesson designed specifically for women with small hands.

Sadly, David Chacko passed away some years ago. I lost a good friend and a real asset. He was a great mentor, and a very good novelist who published a couple dozen books. The plot of one of his novels was stolen and used in a Hollywood movie. That says something about his creativity and the quality of his writing. One of his early publications was a textbook on creative writing which was used for years at the college level by other professors.

The novel I wrote while in the Vermont College program (*After NAM: a Police Story)* did prove useful when I was trying to talk Colonel James Kasler into allowing me to write his biography. I self-published *After NAM: a Police Story* in 1993. (Back then, it was called "desk top publishing" and that was much more expensive than current self-publishing). I never realized any net profit from sales of that book, but I gave Colonel Kasler a copy as proof that I could write something of book length. To that end, my investment of time and money was worth it.

Also, my second master's degree (in writing) did boost my salary a bit during my last few years as a public-school teacher.

By 1999, I was pretty well "burned out" as a teacher. I had bought back my military time early in my teaching career by paying my mandated half into the retirement system, plus the school district's half. All told, that put me just over the 30-year threshold for early retirement.

Coincidentally, (there's that word again) in 1999 the state legislature in Pennsylvania, offered a "no penalty" retirement at 30 years. Ordinarily I would have had to wait another five years for a "no penalty" retirement, or take about a 15% permanent reduction in retirement benefit. Again, the timing was perfect. I took the retirement at thirty years and have never looked back. Contrary to popular myth, teachers in Pennsylvania do not receive 100% of salary in retirement. The formula is 2% per year of service times the number of years served. My 30 years of service times 2% equals 60% of "final average salary." ("Final average salary" is the best three years averaged). If I had 30 years of military service, that military retirement would produce 75% of my "base pay"—a lot richer retirement—plus the military provides "Tri-Care" medical care for their retirees. Teachers pay for their medical insurance in retirement just like most other citizens. I'm not complaining, you understand. I'm just stating the facts.

After I retired in 1999, I went after Colonel Kasler's biography full time. That first year, I put 30,000 miles on my car, running around the country interviewing not only Kasler, but also his son in Florida, his older daughter in Atlanta, Georgia, his younger daughter in Indiana, his cell mate from Hanoi in Tennessee, as well as BGen Robbie Risner when he lived just south of Austin, Texas, among various others.

When I ran out of people to interview, I asked Colonel Kasler if there were anyone else I should talk to. Kasler is totally unlike the persona of Tom Cruise in *Top Gun*. As one of his friends put it, "The words barely slip out of his mouth." He mumbled a name, and I responded, "Who?" He mumbled it again. About the third time he said the name, I responded, "Are you talking about retired Major General Hoyt S. Vandenberg, Junior?" He replied, "Yes, my old friend Sandy Vandenberg."

I said in response, incredulously, "How do I get ahold of him!?

"Here," he said. "I will give you his phone number." So, I drove all the way down to rural Tucson on a separate trip to interview him for two-and-a-half days—on tape, as a house guest.

CHAPTER TWENTY-FIVE

Colonel James Kasler

See Chapter Fifteen for a review of how I crossed paths with Colonel Kasler. He was already a combat veteran of WWII and Korea when I served under his command for that NATO competition at Chaumont Air Base, France. Part way through his three-year tour of duty at Bitburg, he volunteered for duty in Viet Nam. (He was not the only Bitburg pilot to do so). There is a photo in *Tempered Steel* (figure 12) which was taken at Bitburg Air Base in 1965 which shows James Kasler, Bob White, and James McInerny in Air Force "mess dress" uniforms of that era. All three (among several others) later went to Vietnam. James Kasler became the only pilot **ever** to earn **three** awards of the Air Force Cross (which is second only to the Medal of Honor as a combat valor decoration). Bob White and James McInerny each earned a single award of the Air Force Cross in Vietnam. That's **five** future awards of the Air Force Cross in that one photo—which is a rare award for valor in combat. I would also remind that one cannot predict the future. Also in that same photo, all but hidden behind James McInerny, is Hoyt S. Vandenberg, Jr. (and his wife). "Sandy" Vandenberg, Jr. is the son of the second Chief of Staff, USAF. Sandy's wife was the daughter of a Medal of Honor recipient from the Ploesti Raids of WWII, a B-24 pilot. Again, it is a small world, especially in the fighter-pilot community.

As noted, I spent two-and-a-half days as a house guest of Major General Vandenberg, Jr. and his wife while I taped interviews with him for Colonel James Kasler's biography. That was golden. Vandenberg Jr. was a treasure trove of information. He had met everyone from Charles Lindbergh and General Curtis LeMay on down, because they all passed through his father's house. I brought up General Curtis LeMay's name because he commanded the Pacific Air Force during WWII when Kasler served in that theater of operations. Vandenberg, Jr. said, "Oh hell, Curt. I shot trap with him when I was a teenager." That's not bragging. That's just how his life was as son of the second Chief of Staff of the Air Force.

I also got a great quote from Vandenberg, Jr. about the rank-order of the best "dog fighters." He stated that the "three best stick and rudder men in the business were Chuck Yeager, Jim Kasler, and Robbie Risner." (In that order.) That's extremely high praise. Vandenberg, Jr. had flown mock aerial combat with each of them—and many others--in the era before video tapes and all of that electronic technology. Furthermore, Vandenberg, Jr. said that after they landed, Risner reviewed every detail: what he, Risner, had done, and what Vandenberg had done, and then told Vandenberg what he should have done to counter what he, Risner, did. That is the mark of a great teacher and why Risner was the unofficial commander of "Red Flag"—the Air Force equivalent of the Navy's "Top Gun."

When I first arrived at Vandenberg, Jr.'s home, General Vandenburg, Jr. met me at his decorative iron gate. He studied me for a moment, and then said, "I remember your face from Bitburg." I thought to myself, "Why would he remember the face of one lowly airman among the thousands of lowly airmen at Bitburg?" I certainly wasn't an airplane "crew chief" or someone who served the officers in the headquarters building on base. How or when had I crossed his path at Bitburg?

Then, upon entering his house, I found that Mrs. Vandenberg was a collector of antique furniture. I spotted a tall clock across the room and walked up to it to see who the maker was. The clock face said "Daniel Scheit, Sumneytown." I turned to Mrs. Vandenberg and asked, "Do you know where Sumneytown is?" She replied, "No."

"Well," I said. "It's only a few miles east of where I live in Pennsylvania." (Again, a small-world coincidence.) Apparently, it was the only tall clock she owned.

I also spent a long afternoon with Robbie Risner at his home when he lived just south of Austin, Texas. Wow. He had a large trophy room filled with memorabilia from his three-war career, including several examples of the "Risner Trophy," a framed, chrome-plated barrel of a 20mm Gatling gun, and a number of ceremonial swords, etc. that had been presented to him over the years.

At the end of that session, Brigadier General Risner told me the best fighter-ace story I have ever heard. Risner was at a "gathering of aces." To begin with, one must know that the ticket for admission to such an event is that one has to be an "ace." Any war counted, regardless of which side one participated on. So, Risner found himself being queried by a Russian general who had flown in the Korean War—for the "other side." That Russian general wanted to compare notes with Risner to see if they might have "crossed swords" (so to speak) in the skies over Korea. Risner looked the Russian straight in the eyes and stated flatly (with absolutely no animus), "Nope, if we had, you wouldn't be here."

I repeated that story to an aviation writer with Communist leanings. His response was, "Oh that Robbie Risner: that arrogant son-of-a-bitch." I explained, calmly, "You don't understand. Risner knew his F-86: its strengths and its weaknesses. He knew the MiG-15: its strengths and its weaknesses. And he was absolutely certain that if he got into a dog fight with a MiG, he was going to win. End of discussion."

I should also mention that Risner was part Native American. Kasler himself was one-eighth Miami of Ohio Indian—because his great grandmother was captured by that tribe as a child and raised as one of their own—including being given to a brave as his mate. A later treaty led to the return of all the white "captives" - including his great-grandmother and the child she had by her native American mate.

Kasler was also descended from a so-called "Hessian" soldier from the American Revolution. However, Kasler knew none of the details: no name, no dates, no battles, nothing. Over time, I found all of those details.

As I researched on the Internet, I would search for the name "Kasler." Some were dead ends. For instance, one family had *assumed* the name Kasler. However, I eventually found a man in Ohio named Kasler, a genuine descendant of the original Hessian soldier. He was able to supply a family tree for Colonel James Kasler's line of descent.

The first Kasler ancestor to America was Johann Michael Kasler, a 16-year-old boy who was conscripted and subsequently critically wounded at the Battle of Bennington, August 16, 1777. He took a musket ball through the lower leg, breaking both bones. Down on the ground from that wound, he was shot through the chest, point blank, by an angry American colonial. The young Hessian soldier was treated by a Prussian civilian doctor, a volunteer, who later attested that the ball caught both lobes of the young soldier's lungs but (fortunately) missed the heart. One might imagine that when the young Hessian soldier saw it coming, he turned to avoid the shot, which then passed from one side of his chest to the other behind the heart but hit both lungs. At that point, he should have bled to death—literally drowning in his own blood.

Johann Michael Kasler was found, unconscious, but still alive, by an American patriot, one Peter Howe, who got help and carried the young "Hessian" to a house where soldiers were stationed. He was treated there by a volunteer civilian doctor, a Prussian. Then they hauled young Kasler to a colonial American home where the daughter of that family, Suzanna Minkler, nursed the gravely wounded Johann Michael Kasler back to health, then married him. Now, the story becomes a fairy tale—and connects with another coincidence: the modern Colonel James Kasler's older daughter is named Susan, a variant of Suzanna. When the Kasler's named their first-born, they had no knowledge of Suzanna in their family history.

When I took the Hessian soldier story to the people at the Historical Society at the Bennington Museum they got that glazed-over look on their faces and dismissed the story as "a legend." When I came back some time later, they would not even let me into the museum. However, their parting shot was golden: "Have you checked with the Johannes Schwalm Historical Association?" My response was, "I have never heard of them."

They responded, "Well, they are located down there in Pennsylvania." So, when I arrived home, I got on the Internet and checked. They were no longer active, but all of their records were on the Internet.

And there he was: Johann Michael Kasler, where he was born (in a place called Shuttenberg) in what later became Germany—since Germany was not yet a nation until 1871--, the Battle of Bennington, the name of the Prussian doctor who treated him, whom he married, where he was buried, etc.

I was able to locate an article printed in Abbey Maria Hemenway's *Vermont Historical Gazetteer* in 1871 in which Dr. H. H. Reynolds interviewed the Prussian doctor, one Dr. Jacob Roebuck (also spelled Roebeck, Rubach, or Rebeck in various accounts). Dr. Reynolds also attested to interviewing three Hessian soldier survivors of the Battle of Bennington who were treated by the Prussian doctor. Those Hessians were Taring (or Doering) from the Brunswickers, and Row (or Rau), and Michael Kesler (or Kessler) of the Hanau Regiment Erbprinz.

I would note that if one puts an umlaut over the "a" in Kasler, it would be pronounced "Kessler" in German. It's the same name. (John McCain pronounced it that way when he was interviewed by Geraldo on his TV show).

I was also extremely fortunate to unravel the story of the man who saved the wounded Hessian soldier. A colonial woman named Sophronia H. Trowbridge wrote a tribute to her father, Peter Howe, in 1874. Peter Howe was about as common a name as "James Smith" in the modern age—and there was more than one Peter Howe in the historical records who fought at the Battle of Bennington. However, Sophronia Trowbridge is an unusual name. So is Susanna Minckler, the young American woman that Johann Michael Kasler married. (Thank goodness for unusual names.) In Sophronia Trowbridge's tribute to her father, written late in her life, she identified a Hessian soldier whom her father saved. When Johann Michael Kasler passed away, his wife Susanna and two sons moved to Ames Township, Ohio. Coincidentally, they accidentally became neighbors to Peter Howe, the man who had saved Johann Michael Kasler's life. Small world, once again, and a highly improbable coincidence.

With that evidence in hand, I went back to the Historical Society at the Bennington Museum. They had to admit that it was not a fairy tale. Subsequently they published my article in their journal, *The Walloomsack Review,* Volume 9-September 2012—pp 6-11.

In their defense, let me say that the people in that museum/historical society had been frequently presented with "tall tales." For instance, someone would show up with a musket and claim that it was carried by their great, great, great grandfather at the Battle of Bennington. Inevitably, the story never panned out. The musket was from a later era, or was a forgery, etc.

By the way, Michael Kasler was born in 1761. Since he was only 16 for the Battle of Bennington (in 1777), he never died until 1839. That brings the story much closer to the modern day than one might imagine.

One might also question why an American girl would marry a "Hessian soldier." Consider that this was a very large country back then, with not nearly enough men to work the land or to marry young women. Former Hessian soldiers had only to sign a document promising never to bear arms against this nation again, and they were free to live out their lives in America, to own land, and to marry. Very few of them returned to their former home land—where they were "serfs"—a station in life little better than that of a slave. They didn't volunteer to come to America. They certainly did not sign up for a pay check. They were sent by the archduke or prince or whoever was the ruler of that particular small kingdom, of which "Hesse" was one of many such. That archduke, etc. received a rental fee for the use of the serf, and if that serf were to be killed, that archduke, etc., was paid a bounty. It was a "win-win" for the archduke, but "lose-lose" for the serf.

CHAPTER TWENTY-SIX

Getting Published

Getting published is a tough assignment. One needs a track record. That is a catch twenty-two. One needs a track record to get published, but it takes publications to have a track record. A self-publication does not count.

Meanwhile, Kasler's son retired from the Navy, and with time on his hands, he set about finding an editor to help get my manuscript in better shape. (I was having trouble with how to present the "other voices" in my first person, oral history version of Kasler's biography. I knew that approach could work because that is what Yeager did with his first-person oral history).

Kasler's son interviewed several candidates for the job of editor for what became *Tempered Steel*. However, one after another, they would come on board, take a look at the manuscript, and withdraw from the project. When Lt Col Perry Luckett, PhD, came along, I thought he was just one more such editing candidate. However, he certainly was not.

He was a "full professor" at the Air Force Academy who also taught at the University of Colorado. He had published one book: *A Bio-bibliography of Charles Lindbergh*. Contrary to my assumption that Luckett was one of Kasler's son's recruits as an editor of my first-person, oral history manuscript of Colonel Kasler's biography, Perry Luckett came on board as an editor through an entirely different avenue. (See page X of the Acknowledgments in *Tempered Steel* as it was published by Potomac Books, Inc.) It states:

"Col Frederick Kiley read the entire manuscript early and late, then responded [to Perry Luckett—not to me] with excellent suggestions concerning the structure, section headings, and areas for improvement. Kiley's book, *Honor Bound: American Prisoners of War in Southeast Asia, 1961-1973 (with Stuart I. Rochester)* was also a vital source of information about the prisoner-of-war experience in Southeast Asia. Equally important was his experience with the publishing process: **he was largely responsible for finding Brassey's, Inc. as the publisher of this biography."** (These are the words of Perry Luckett—lieutenant colonel and PhD. Unfortunately, I was not "in the loop," so to speak, during that discussion.)

Let me say, up front, that I will always be grateful to Perry Luckett and the two men (Kiley and Rochester) who were responsible for finding Brassey's, Inc. as the original publisher of *Tempered Steel.* However, my original manuscript was oral history written in the first person—from the horse's mouth of the men I interviewed. That was not an idle decision on my part. My models were Robert L. Scott's *God Is My Co-Pilot* and Chuck Yeager's autobiography *Yeager.* Both were done with a recording device—in the first person. Scott's book was so early that he used a wire recorder. (Mylar tape had not yet been invented.) Yeager used a tape recorder. I might note that both of those books were great financial successes. Robert L. Scott was still selling copies of *God Is My Co-pilot* until he died.

Chuck Yeager, shortsightedly, passed the copyright of *Yeager* to his children, thinking the book would have only modest sales. His offspring made millions in royalties from the book's sales.

Since I still retain the copyright as author of my original, unpublished, first-person manuscript of Kasler's biography, I intend to self-publish that separately.

Incidentally, I have the opinion of two professionals (one retired Air Force major general and one professional videographer) who have read both versions. They both prefer my "oral history" version of Kasler's story. I will let my readers decide.

Charles L. Byler, co-author of *Tempered Steel*
and author of
Colonel James H, Kasler: An Oral History

APPENDIX

The following short story was originally titled "Sam Vater's Ordeal." I, Charles L. Byler, wrote it while I was an undergrad student at Slippery Rock State College (now Slippery Rock University). That short story won "first prize for fiction" in the Spring of 1969: Volume V, number 2 of, the literary magazine of Slippery Rock State College. The faculty advisors of *Ginger Hill* were the late Robert Flanagan (a novelist) and Brendan Galvin (a poet, now regarded as the finest American poet alive and still writing today).

I would like to change that title to *Death of a Thunderchief.* The original title was a tip-of-the-hat to William Faulkner and his character "Sam Fathers," a black man in a story about a bear and a dog named Lion. I will make some minor editing changes to my short story.

As noted in the text of my memoir *Gun Plumber,* this short story is based on my eyewitness account of the fatal crash on take-off of an F-105 Thunderchief fighter bomber during my first "Operational Readiness Inspection" at Bitburg Air Base in Germany in 1963, killing the pilot who was coincidentally the only black pilot in 36th Wing at that time. That fact may be purely coincidental. However, as another black man once told me, "I had to be 150% better than a white man to get the same recognition." I may speculate that the black pilot's practice of raising the landing gear handle to the "up" position just after he started his take-off roll, may have been just such an effort to be 150% better than a white pilot in order to be noticed. At any rate, that practice put him in an unrecoverable situation on that fateful day when things went terribly wrong.

There is also another contributing factor here: that same aircraft had been written up by that same pilot in the morning flight that day. There had been a "failure to drop" the bomb on that morning mission. The pilot wrote up the malfunction, then was assigned to that same aircraft for the afternoon mission. The malfunction had been checked out by ground crews as "ground checked okay." In other words, no problem was uncovered. Perhaps the pilot had screwed up. Something as simple as forgetting to push in the "master armament circuit breaker" could have produced the failure to drop. The pilot's reputation was on the line. A second "failure to drop" would not look good on his evaluation report—especially during an "Operational Readiness Inspection."

I have another theory based on a strange coincidence later as a weapons mechanic. Unfortunately, at this late date there is no way to test my theory.

Another F-105 was written up for a "failure to drop" and my crew was sent to the flight line to check out that aircraft. Our check-out failed to detect any problem. Then, I happened to notice that the "ground override switch" in the nose wheel of the F-105 had not been activated. That means the weapons systems should have been "cold"—and yet we were getting indication that the system was "hot" or "live."

So, I went to the nose wheel compartment, removed the lid on the safety override box, and pulled the safety switch out. That should have made the armament systems "hot"—but now those systems were "cold." What gives? It was later determined that the ground safety switch had been wired backwards at the factory. The aircraft was "hot" on the ground and "safe" in the air. No wonder, the bomb would not drop. On the other hand, this was an "accident waiting to happen" on the ground. We will never know the truth. All of the relevant evidence is long gone.

However, most of our F-105s at Bitburg had been manufactured in 1960. However, that malfunctioning F-105 had been manufactured in 1961. The F-105 which crashed fatally on take-off had also been manufactured in 1961. Coincidence?

On one of those TV shows about renovating homes, a technician noted that when one finds one deviation from the building codes in a home, there are likely other deviations from the building codes to be found in that same home. If one aircraft was wired backwards at the factory, then there would likely be another such mistake, probably made by the same person at the factory.

"The Death of a Thunderchief"
A short story by Charles L. Byler

The air base alert siren wailed at 5 a.m.

"Son-uv-a-bitch" muttered Airman Braun dropping his bare feet onto the cold tile floor and feeling around in the dark for a cigarette. Now the horn inside the barracks began bleating. Grey, Braun's roommate, was jarred awake by this second, louder blast and leaped out of bed. The moment of fear dissolved into a string of curses. Turning on a light he said, "Why the hell we the only ones with a damn siren in our barracks? Not even the damn sky cops gotta put up with this fuggin' noise!"

"'Cause we're weapon loaders," drolled Braun. "Those planes can't go no place until we get down there and hang some bombs on 'em."

In apartment 6B of the base housing unit fifteen, fighter pilot Captain Sam Vater's wife reached over and shook her husband without opening her eyes. "Get up," she murmured. "The siren's blowing."

Yeh. Might a known, thought Vater, stirring. *Wish they'd a waited till daybreak at least.*

Then out loud to his wife, "Don't bother getting up. I'll just get a glass of orange juice."

By the time he had gotten his flight suit on and had drunk the juice, the sirens had stopped and his wife was falling asleep again. When he got to his car, there was already a solid line of headlights racing for the flight line. He nosed into the traffic and sped for the flight shack. Rolling up in front of the 23rd operations building, he saw several yellow bomb trailers in a convoy crawling towards the long line of planes. Each trailer carried two concrete-filled practice bombs and was towed by a pick-up truck with a blue flashing light and an armed guard. A white-striped blue air police vehicle led the stately procession.

Man, those weapons people are really quick, he thought. *They'll have every bird on the line loaded before dawn. Fine lotta good that'll do – can't use the bombing range till the sun's up.*

An operational readiness referee replete with an armband, clipboard, and heavily-starched fatigues was poised pen-in-hand (and sickeningly bright-eyed) next to the alert sign-in board. Vater gave a matter-of-fact "Hi" as he picked up the red grease pencil and printed his name on the big plastic sheet. Finding a chair in the back corner of the room, he plopped down and thought, "*Here we go again. Take X number of planes with X number of bombs down to Suippes Range and drop that X number on the same damned target. That range is beginning to look as familiar as my back yard.*

Five years now flying this same hog: two years on this same base. Why the hell don't I just quit? Could make more flying civilian at not half the risk. Wh..."

"Tench hutt!" yelled a lieutenant in the front row as the squadron commander, Colonel Greene, strode in.

"Be seated, gentlemen," the colonel said and everyone slouched back into his chair. The briefing was just that---brief. They had been through this too many times. All old hat.

"Yep", thought Vater, *"We've had six practice alerts in the past three months preparing for this O.R.I. If the damned planes don't break we're sure to pass with flying colors – and the wing commander – bless his bloody soul--* [our actual commander's name was "Blood"] *will make general and get a new job at 17th Air Force and we'll get a new colonel straight out of the Pentagon. Then the whole mess will all start over again. Glad I've only got one more round to go before I get the hell out of here. What a base – one general per year – just like clock work. Doesn't seem right – We do all the work and* **he** *makes the rank!"*

Twenty minutes later, the dawn was breaking as Vater jumped down from the still-moving Metro van and he walked toward aircraft #128.

"Hi," he said to the aircraft crew chief as he handed him his helmet and bag. The crew chief climbed the cockpit ladder and deposited the helmet and bag in the cockpit as Vater began checking the aircraft forms.

Everything's OK here, he thought. She's just come through a fifty-hour inspection, too. He stuck the forms back into the notch in the ladder and began a leisurely walk-around inspection: kicking the tires and checking safety pins. He signed the forms and climbed the ladder into the cockpit.

Swiftly and methodically, he ran through the checks, flipping toggle switches, turning knobs, pressing buttons and tabs, and checking circuit breakers. When all his hoses and lanyards were connected, the crew chief gave him the wind-up signal. Vater pressed the button—the start cartridge ignited with a hiss and acrid black-powder smoke filled the air as the turbine wheels began to spin, whining as they gained speed. Now he lit the fire in the jet engine and brought the throttle quadrant to the idle position. Working with the crew chief through hand signals, he flexed the flight control surfaces and speed boards.

"OK. Pull the safety pins and chocks." Under the crew chief's direction, Vater eased out of the parking space onto the taxi way. He returned the crew chief's salute and rolled towards the end of the runway. *Right on time,* he thought. *Just like clockwork.*

But when he got to the run-up area, one plane was turning around and heading back.

What the hell! Don't tell me we've got an abort already!

Sure enough. He heard the radio confirmation. Major White couldn't get full power out of aircraft #484.

Vater moved up on the ramp and did his engine check. *Flaps down.*

Then he rolled out for take-off. Final clearance. Advance throttle. Release brakes. Full military power.

As the aircraft rolled, Vater moved the throttle quadrant to the left to light the after-burner. Now the ground-speed indicator rose rapidly as the thousand-foot marker flashed by. 90 knots—1500 feet—2000 feet—120 knots—2500 feet—150 knots—Vater jerked the landing gear handle to the "up" position—prematurely so the landing gear would retract smartly when the weight of the jet fighter plane was lifted off the squat switches in the landing gear. 150 knots. Vater eased the slick back and the big fighter plane lifted off. The landing gear did indeed retract smartly. A clean take off.

Navigating by map and radar, Vater sped towards the bombing range in France. Immediately, he began setting up switches for the bomb run. He wanted to be ahead of the game. No last-minute hustle to catch up.

The visual target identification point appeared and he started his drop pattern, watching the timer and toss bomb computer.

At the drop point, he pressed the "pickle button" to release the bomb---but nothing happened. No drop.

"Son of a bitch!" he muttered to himself. *No second chance today.*

Quickly, he rechecked his switches. *They're all AOK ! What the hell!*

Pull the master armament circuit breaker. No use having an accidental drop now!

Back at the air base, he angrily wrote up the malfunction in his aircraft forms and, dejected, walked back to the flight shack, carrying his parachute slung over one shoulder.

Before long, the group of pilots were assembled and the colonel strode in—clearly hopping mad.

"We're coming unglued at the seams!" He barked. "We haven't done this poorly since last summer. One abort—one failure to drop—and one total miss! And we're not halfway through this exercise yet!" His voice rose progressively towards the end of his sentence. After a few more curt remarks, the colonel stomped out the door and Major Blanche took over the debriefing session.

Son-uv-a-bitch! thought Vater. *The old bird is laying the blame on us! Hell! What can a pilot do about a lousy broke-dick bird? Weiss says the TBC screwed up on him. The miss wasn't his fault. White's engine flaked out on him. And I know my malfunction wasn't between the head sets.*

That's a chicken-shit colonel for you! Could care less who's at fault. Just sweats losing his almighty command position and promotion! Dammit! I'll fix him! Next time I'll make a manual, timer drop! No safe-separation time! No TBC! Nothing! To hell with it all!

The crew chief could see the fire in Captain Vater's eyes when he stepped down from the van later that day for the second mission.

How do I rate? Thought Vater. *The same broke-dick airplane for the second mission of the day.*

He grabbed up the aircraft forms and flipped them open to his own write up. "Ground checked OK in accordance with T.O. #......"

Son of a bitch! Bet they didn't fix a damn thing! Damn lazy incompetents!

Vater's walk around was very brief. After all, he had ground-checked this same bird in the morning. But he kicked the tires twice as hard as usual and shook the wing tanks vehemently.

His check-list run down, however, was very meticulous in the cockpit. *There will be no slip-ups this time.*

Out on the runway, he watched Major White lift off and then began his take-off roll. Vater lit the burner and watched the 1000 -foot marker flash by.

"Hey Braun!" yelled Airman Grey. "Look at that!" Braun had already looked up and dropped his tool bag as he heard that after-burner explode. He scrambled up onto the roof of a pick-up truck to get a better view. The after-burner flame was twice as long as usual—nearly thirty feet long.

"She's blowing up! yelled Braun. "Shut it down! Shut it down!"

Pieces of metal began blowing out of the cone of the exhaust and dropping onto the runway. Vater couldn't see the tail cone of his plane or the pieces of metal being ejected out his exhaust. The sound of the after-burner sounded normal inside his tightly closed cockpit.

To Braun, Grey and others outside the plane, the sound of the jet engine had an abnormal rattling roar—like a dragon with a sore throat. And there was that abnormally long flame coming from the exhaust cone. The men in the control tower could see that abnormally long flame.

Vater heard a crackly voice calling "Abort! Abort!" in his head set, but he was halfway down the runway and had already started to rotate for lift off.

When his main landing gear lifted off the runway, the landing gear immediately started retracting. When the hydraulic system begins to operate, power is robbed from the jet engine—which was already losing power from the damaged after-burner. The F-105 did not have sufficient power to support flight. The big fighter plane settled back onto the runway. The two 450-gallon wing drop tanks scraped on the concrete runway, igniting both instantly in two puffs of flame and black smoke. The belly of the fighter plane skidded on the runway, along with the tail section of the plane.

All winged aircraft rotate on a lateral axis which runs through the wings. The tail must drop in order to lift the nose. With the tail dragging, there was no way to lift the nose. Vater could have deployed the tail hook, but that would have been futile. The weight and speed of the jet fighter would simply have torn the tail hook off. The other barrier on the end of the runway was designed to snag the landing gear—which had been retracted.

Getting rid of the one-ton dummy bomb in the bomb bay would have reduced weight for lift off, but there simply was not enough time. The "panic button" would send ejection voltage to all the external stations, but not to whatever was attached to the rack in the bomb bay—whether it was a live nuke or a dummy. That weapons station had to be activated by the SWESS (Special Weapons Emergency Separation System). There was simply not enough time to activate that system.

Vater could have used his ejection seat, but he did not have the altitude needed for that to save him. He was along for the ride, a very short and fatal one.

When any winged aircraft stalls, one wing always drops first. In this case, it was the right wing, which caught earth off the end of the runway. That tore the right wing off the F-105 and left that wing standing upright on the leading edge of the wing with the bottom of the aluminum wing and the large, blue, block-lettered "USAF" exposed like a billboard. The F-105, like all Republic fighters, was a stoutly-built aircraft. The shock of ripping the right wing off the fuselage, slammed Captain Vater's body into his safety harness so violently that his spine was broken in his neck and the spinal cord in his neck was severed by the weight of his head and helmet. He died instantly.

The fighter plane slewed to the right and plowed into a stand of closely-planted trees.

Braun, Grey, and all of the others within earshot on the airbase heard the WHUMP of the plane hitting the trees, followed by an echo of the impact—and then a whine of the turbine blades in the jet engine winding down to silence.

In the nearby base housing area, dozens of women also heard those sounds—and wondered which one of them was now a widow.

PHOTOS

Cadet Warrant Officer Charles Byler receiving marksmanship award at Penn State ROTC

Department of the Air Force

CERTIFICATE OF TRAINING

This is to certify that

[illegible]

has satisfactorily completed the

[illegible]

Given by

[illegible]

James Salter and James Kasler flew the F-86 in the same squadron in Korea. Photo courtesy of James Salter.

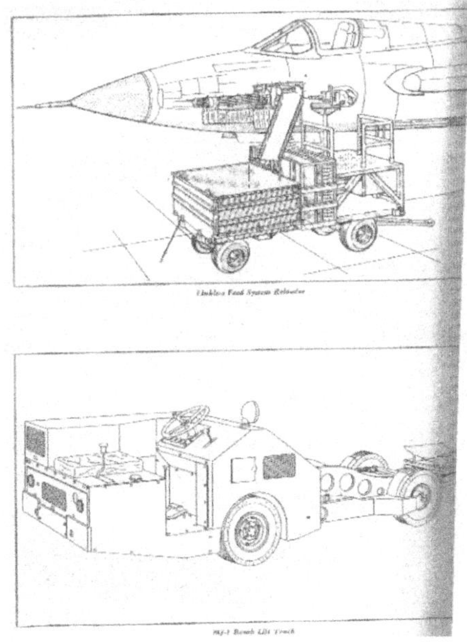

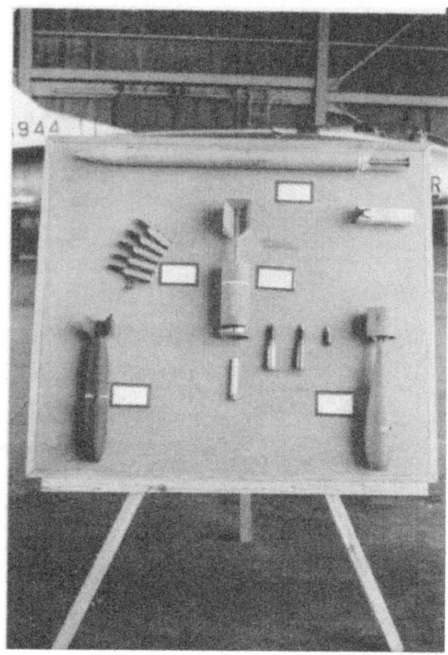

Practice munitions for the F-105

Dummy nuke

Kasler Memorial

Highlights of Kasler's career

Chuck's dirt track racecar

James Clark at Zandvoort, Netherlands – the 1965 Grand Prix (which he won). This photo took first place in a United States Air Force in Europe photo contest.

Ken Smith - one of my writing professors at Slippery Rock, noted as the best English poet in the latter half of the 20th century

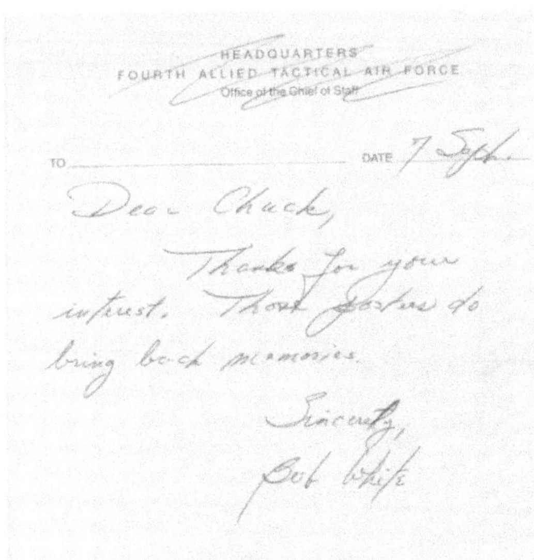

HEADQUARTERS
FOURTH ALLIED TACTICAL AIR FORCE
Office of the Chief of Staff

TO DATE 7 Sept.

Dear Chuck,

Thanks for your interest. Those photos do bring back memories.

Sincerely,
Bob White

Bob White - X-15 astronaut and recipient of the Air Force Cross for Valor flying the F-105 in combat

My grandfather's home after it was arson burned

The house I was raised in on Reynolds Street - again, arson burned

All photos on pages 116-120 taken by Chuck Byler.

ACKNOWLEDGMENTS

"No man is an island entire of itself."

I would like to acknowledge Brendan Galvin, PHD in writing, as well as my poetry and writing professor at Slippery Rock State College. He was the first person to tell me that I was a writer. Another professor at Slippery Rock who had a positive influence on my writing ability was British-born poet Ken Smith (who is today regarded as the finest English poet in the latter half of the 20th Century).

The late David Chacko was my "field faculty advisor" when I was writing a novel as my "major product" for my Master of Arts in Writing degree from Vermont College when it belonged to Norwich University. David Chacko authored a college-level textbook on writing, as well as more than 20 novels. As proof of his talent, I would note that the entire plot of one of his novels was stolen for use in a Hollywood movie without using a single phrase of his written text--to avoid having to pay him any royalties. (Boo! Hiss!)

Extra-special note belongs to Lisa Gery who was one of my students when I was teaching public school English. She was the only student during my 27 years as a teacher who handed me the manuscript of a novel which she had written. I knew then that she was a writer. She went on to self-publish *The Persistence of Vision* and to form her own publishing company: Underdog Publications.

More recently, Lisa did the formatting of my memoir: Gun Plumber, the Auto Biography of a Former F-105 Weapons Mechanic/ Weapons Loader. (Since I was born before the advent of TV and computers, her assistance has been invaluable.) I date myself by noting that when I went to Penn State's Main Campus in 1961 there was only one computer on campus. It filled a brick building and the vacuum tubes heated the whole building in the cold mountains of central Pennsylvania.

At Slippery Rock I collected the data for my first master's degree on IBM punch cards. Years later, I realized that there were no longer any

IBM machines to read those cards. Reluctantly, I put those 2600 IBM cards in a recycling box.

Last, but certainly not least, I would also like to acknowledge Colonel W. Howard Plunkett, a former F-105 maintenance officer for the F-105 THUNDERCHIEF who has published five books on the F-105 as well as *Pack Six Blues-- The Diary of an F-105 Thunderchief Pilot.* It was retired Colonel Howard Plunkett who noted that one perspective on the F-105 that had not been written about was that of a Weapons Loader/Weapons Mechanic.

The Persistence of Vision
Three lives. Three secrets. One tragedy.
Lisa Gery

TEMPERED
STEEL

Perry D. Luckett
and Charles L. Byler

The Three Wars of Triple Air Force Cross Winner
JIM KASLER
Foreword by James Salter, author of *The Hunters*

The first printing of the first edition of *Tempered Steel*

MILITARY HISTORY/BIOGRAPHY

"I was privileged to serve in the company of heroes . . . like Jim Kasler. They were the ones who sustained me."

—SEN. JOHN MCCAIN

"Jim Kasler is an indestructible man who persevered over death."

—VICE ADM. JAMES STOCKDALE, USN (RET.)

"You couldn't pick a better person to read about than Jim Kasler. He's a paragon of what a fighting airman is all about. He didn't break the sound barrier first or go to the moon and jump up and down. He's just the best at his trade that anyone could be, and he is a complex individual whose characteristics all came together when he needed them to produce what he did. I consider him a genuine American hero, and I hope our Air Force has a few like him in the future when we need them."

—MAJ. GEN. HOYT SANFORD VANDENBERG, JR., USAF (RET.)

ISBN 1-57488-834-X
90000
9 781574 888348

Made in the USA
Middletown, DE
23 August 2024

59532222R00081